IET MANAGEMENT OF TECHNOLOGY SERIES 28

Lean Product Development

Other volumes in this series:

Volume 15	**Forecasting for technologists and engineers: a practical guide for better decisions** B.C. Twiss
Volume 17	**How to communicate in business** D.J. Silk
Volume 18	**Designing businesses: how to develop and lead a high technology company** G. Young
Volume 19	**Continuing professional development: a practical approach** J. Lorriman
Volume 20	**Skills development for engineers: innovative model for advanced learning in the workplace** K.L. Hoag
Volume 21	**Developing effective engineering leadership** R.E. Morrison and C.W. Ericsson
Volume 22	**Intellectual property rights for engineers, 2nd edition** V. Irish
Volume 23	**Demystifying marketing: a guide to the fundamentals for engineers** P. Forsyth
Volume 24	**The art of successful business communication** P. Forsyth
Volume 25	**Effective team leadership for engineers** P. Wellington
Volume 27	**How to build successful business relationships** F. Kay

Lean Product Development

A manager's guide

Colin Mynott

The Institution of Engineering and Technology

Published by The Institution of Engineering and Technology, London, United Kingdom

The Institution of Engineering and Technology is registered as a Charity in England & Wales (no. 211014) and Scotland (no. SC038698).

First published 2000 (0-9538779-0-6)
Revised 2012 (978-1-84919-671-0)

The Institution of Engineering and Technology
Michael Faraday House
Six Hills Way, Stevenage
Herts, SG1 2AY, United Kingdom

www.theiet.org

British Library Cataloguing in Publication Data
A catalogue record for this product is available from the British Library

ISBN 978-1-84919-671-0 (hardback)
ISBN 978-1-84919-672-7 (PDF)

Typeset in India by MPS Limited
Printed in the UK by CPI Group (UK) Ltd, Croydon

Contents

Preface

A manager's guide to organising, running and controlling the business process of developing products

> Our job is to give the customer, on time and on cost, not what they want, but what they never dreamed they wanted; so that when they get it, they recognise it as something they had always wanted – Sir Denys Lasdun

This book is about how you manage the business process of developing products from strategy through design, to testing and service feedback. You can apply it to manufactured and service products, whether completely new or just a minor change.

It's not an instruction manual on tools and techniques. These are explained in *Chapter 11*; *Chapter 12* gives references on their detail.

The book is set out in the order of the tasks you need to tackle. The map (next page) enables easy reference once you've read it. The book can be folded flat open and stored in your A4 files.

'Lean' applied to development

> There are few business problems that can not be fixed by introducing good new products – Carlos Ghosn, Renault/Nissan

The word 'Lean' was coined to mean Toyota Production System (TPS) methodology, developed to identify and eliminate manufacturing waste.

In Toyota, Ohno and Shingo defined the following seven principal manufacturing wastes. You can think of equivalents in developing products:

1. **Overproduction:** over-complex solutions from poor concepts; chopping up the task and letting each department duplicate what others are doing.
2. **Waiting:** for other departments to process tasks, for unnecessary high-level approval; queuing to use an overloaded facility or specialist staff time.
3. **Transporting:** moving bits of product development task to separate departments and then having to fetch them back, analyse and collate their input.
4. **Over-processing:** requiring too many approvals; preparing reports that aren't used; creating features the customer doesn't want.

5. **Inventory:** poorly organized projects that take too long to complete; working on projects that are not commercially viable.
6. **Wasted motion:** searching for information that has not been captured or logically pigeon-holed, often across departmental boundaries.
7. **Rework and defects:** changes to the product, or the way it is to be manufactured, from not considering potential problems early enough.

> 80% of business failures can be traced to the way in which products are developed. But it doesn't depend just on the process manual; company culture is equally important. One doesn't work without the other.

But that's not the whole story; some product development (PD) wastes don't have an exact equivalent.

The main problem is that while you can see manufacturing waste, most PD waste is invisible. It sits in hard drives and (overworked) engineers' brains. The most common wastes are not developing what the customer wants and not offering a sufficiently appealing product. And most companies repeat costly mistakes and take far too long because of expensive rework loops.

Unfortunately, value stream mapping doesn't reveal this until you've already wasted the time and cost. And removing waste from a lousy process won't produce a good one. And it certainly won't create Toyota's learning-based approach or the culture of its development environment.

> If Toyota's success was the result only of replicable techniques and methods, many more companies would be as successful. But they're not.

Toyota's PD success is based on their continuous development of culture, knowledge and hardware. Their store of thoroughly developed background knowledge enables them to apply it rapidly to new products. This is discussed at the close of phase 4.

The book explains how you can avoid waste and revise your culture.

Success lies not only in the grand plan but principally in the detail. And only support from the CEO will generate and enforce the culture. It's your culture that governs how you operate, not your procedure manual!

About the author

Dr Colin Mynott is a Chartered Engineer, a European Engineer, and a past Fellow of the Institute of Directors. A materials technologist by training, he graduated from the London Imperial College of Science and Technology and, for his doctorate, from Cambridge University in 1963.

He has extensive hands-on experience of manufacturing systems and of developing products, from devising strategy through to manufacturing engineering and start-up.

He worked in and managed manufacturing companies that developed their own products. His career started in the automotive industry in manufacturing systems' engineering at the British Motor Corporation; he then became The Materials Engineer for Chrysler UK. Then followed a 5-year spell in international management consultancy. Following that, from 1976, he successively ran medium-sized manufacturing companies in the automotive sector. In a 3-year sabbatical from industry, in 1990 he was asked to organise the UK Design Council's services to manufacturing industry as Industry Director, primarily to improve UK manufacturers' product development capability.

As part of his manufacturing career, he started his own manufacturing firm with two partners in 1973, initially as non-executive chairman and, from 1987, as joint chief executive until its sale in 1989. It continues today as a market leader in its field.

He formed his present company in 1994 to disseminate best practice to industry on all aspects of product development, how it should be organised, its financial impact and management. He is joint author of a number of publications on product development and continuously researches best practice.

When requested, he organises and runs workshops, seminars and conferences on product development to show companies how, through their own efforts, they can improve their performance. And in national programmes with UK institutions, for example with the Royal Academy of Engineering in a UK-wide programme to improve the product development capability of UK manufacturers; and running workshops with the Institute of Mechanical Engineers.

He helps companies with their product development strategy and its operation; for example, with a leading Japanese car-maker in a 3-year programme to help improve the product development effectiveness of their European first-tier suppliers.

In-house workshops

When asked (his contact detail is pd@mynott.com), he works with and facilitates companies, at board and senior management level. The aim is to enable companies to develop products with greater customer appeal in the least time and cost by enhancing their product development culture and process. His in-house workshops have enabled a number of US and UK companies to revolutionise their product development performance and capability. This book sets out the basis; workshops put the culture and process firmly in place across the whole management team.

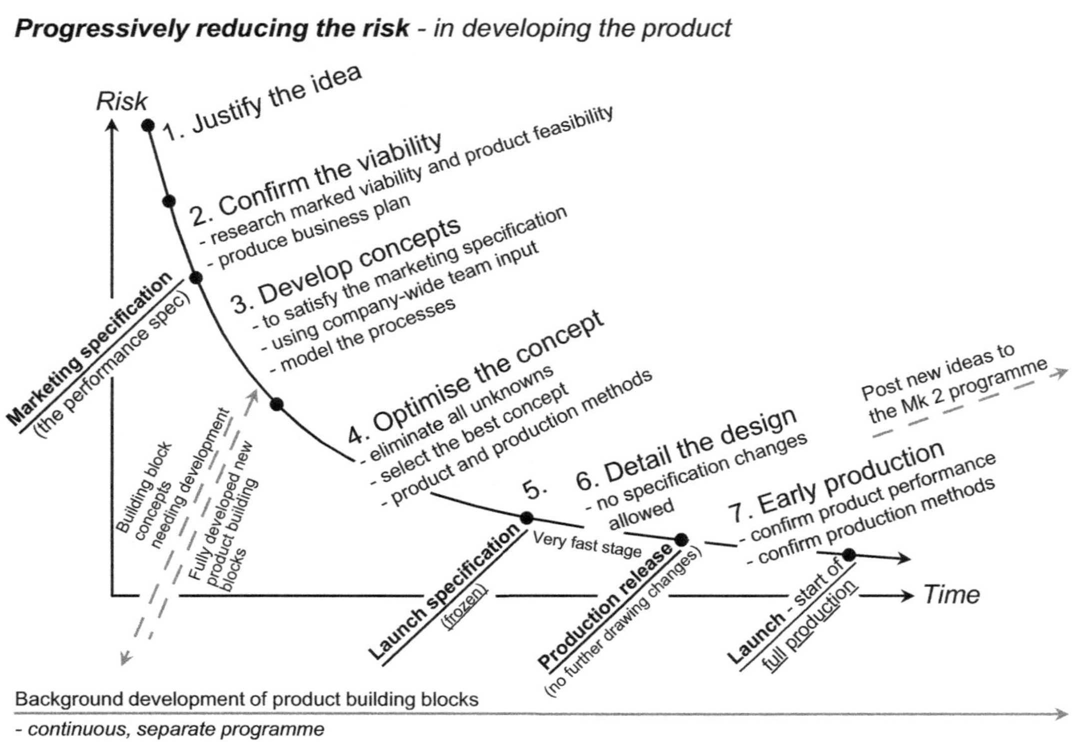
Progressively reducing the risk - in developing the product
Risk
1. Justify the idea
2. Confirm the viability
- research marked viability and product feasibility
- produce business plan
3. Develop concepts
- to satisfy the marketing specification
- using company-wide team input
- model the processes
4. Optimise the concept
- eliminate all unknowns
- select the best concept
- product and production methods
5.
6. Detail the design
- no specification changes allowed
7. Early production
- confirm product performance
- confirm production methods
Post new ideas to the Mk 2 programme
Marketing specification
(the performance spec)
Building block concepts needing development
Fully developed new product building blocks
Launch specification
(frozen)
Very fast stage
Production release
(no further drawing changes)
Launch - start of full production
Time
Background development of product building blocks
- continuous, separate programme

The 7 phases and reviews:

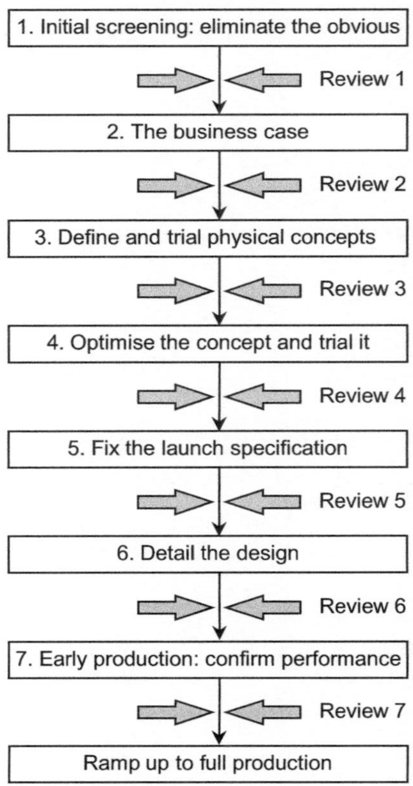

Phase 1. ***Screen the potential project's viability:*** the initiator passes it to a Board-level project champion; using internal company knowledge, they do a fast assessment of fit to company strategy, risk, apparent technical,market and financial viability; Board approves to proceed to Phase 2 or kill.

Review 1. ***Is the idea worth even considering?***

Phase 2. ***Research customer requirements, technical feasibility and costs in detail:*** appoint the project manager (and first team members); the team does detail market research; estimates costs, potential financial return and risk; generate ***the marketing specification*** and the ***project business plan***.

Review 2. ***Is there a quantifiable business case?***

Phase 3. ***Expand the team (from all functions) to generate several concepts:*** evaluate technology, production and design feasibility of each against the marketing specification, their risk and cost. Generate models; trial and quantify unproven elements. Send potentially good but unproven building blocks and ideas to the long-term programme for future products.

Review 3. ***Are there enough feasible concepts?***

Phase 4. ***Generate a better concept from the best features of each:*** evaluate how the features of each concept satisfy the marketing specification; combine the best features to produce a better concept; simulate and make test models; debug interface problems; trial performance; generate a draft launch specification and production engineering specification for the production process.

Phase 4. ***Is there a suitable optimum concept?***

Phase 5. ***Freeze and prepare the detail launch specification and production process specification.***

Review 5. ***Is the launch specification correct?***

Phase 6. ***Detail the design; install and commission the plant:*** issue production release drawings (instructions); install and commission the plant.

Review 6. ***Are the detail designs complete; is the plant commissioned?***

Phase 7. ***Start-up and run the plant; trial early production:*** run early production off production tools; confirm performance and cost against the launch specification; use early products for approval and legislative tests; use the trials to rectify final concerns whether product or process. Check production supplies are organised.

Review 7. ***Are confirmation trials complete; is all ready for production?***

Chapter 1
Introduction

What is the economic purpose of product development? To develop products faster at less cost will not happen without a particular methodology. What kind of companies will benefit from following such a guide?

1.1 What product development is all about

1.1.1 The purpose of product development

I worked in, managed and ran manufacturing companies for three decades. I had always wanted to discover what made the leading companies so successful. How did they achieve it: what particular attributes did they have that less successful ones didn't have? Was it management or process or what else? And was it transferable – could anyone duplicate their success?

A sabbatical in the 1990s enabled me to organise researching the matter in depth. Having been concerned with doing it for so long, I was looking for practical factors that I could apply. And I was seeking root causes, not just symptoms. So we first examined how countries rather than companies became prosperous.

> Product development is not an altruistic activity: the only reason you develop a product is to create more profit than if you hadn't done so.

I was curious to understand the link between a country's prosperity and the performance of its companies, in the manufacturing, service and public sectors. The research soon showed (and maybe I should not have been surprised) that the wealth these countries create is initiated, not by their service industries, but by their manufacturers' product development activities – how well they do it and exploit the results.

It drives their economies. Starting with activity at local level in regions and districts, this catalyses activity further afield and the whole percolates upward to create wealth at national level. In some cases governmental initiatives catalyse this activity. We have seen these, for example, in newly industrialising nations. This has been fascinatingly documented by Ira Magaziner in *the Silent War*, which catalogues the sometimes globally catastrophic effect that such governmental influence can create – catastrophic to complacent nations, that is. (ISBN 0-679-72827-9; essential reading for aspiring politicians.)

It became clear that developing products has always been a key driver of creating wealth. Its purpose is to create profit through satisfying what customers want. And to do that it needs to be driven by common sense operating economics.

The wealth countries create is driven by their manufacturers' product development activities.

1.2 The discovery programme

My three decades of running small to mid-size manufacturers that developed their own products provided a good background to research how the rest of the world behaves.

Starting in the 1990s, with colleagues I started to analyse in depth how more than 70 leading manufacturers went about organising, running and controlling the development of their products. I was able to examine the practicalities of the product development task in minute detail. And not being an academic or consultant, the practical learning points were what interested me most because I could apply them in my own company.

This provided interesting experience. So I helped initiate a national programme with the Royal Academy of Engineering, the engineer's equivalent to the scientist's Royal Society. I developed a simple self-audit tool to allow the company boards of more than 600 product developers, United Kingdom and international, to audit their product development capability. We used about a dozen manufacturing greybeards (me included) to set up the audits. In a return visit, we discussed the audit results with each company, disseminated the lessons learned in previous analyses and stimulated each company to set out a simple improvement programme.

At the same time, following running a few public workshops to disseminate these findings, the occasional company requested in-house workshops to discover how they could apply the lessons in their own companies. This included three years work with Nissan UK to help their suppliers to mesh their product development programmes with Nissan's. This was a key requirement because around 70% of a vehicle is designed by the suppliers, in a myriad of detail component parts from seats to fasteners.

1.3 … and the discoveries …

All this provided a huge database of current practice. And surprisingly, the fundamentals had not changed for decades. Nor were most of the lessons learned taught anywhere.

The obvious questions are: do companies of all sizes in different manufacturing sectors experience the same problems? What methodologies do they use? Is there a good practice operating philosophy equivalent to that set out by Shingo (footnote 1 below) for manufacturing? The answer, after a lot of analysis was yes, there is. We had been looking at the scattered parts of it for more than a decade.

There is a way of developing products that achieves the best result in the least time, more competently and completely and at the lowest cost. But no two projects have identical requirements. You plan each one according to its needs; this book shows how you do so. And the detail of the methodology is continually evolving. In this manual I describe how you go about doing it according to the current state of the art in the 21st century.

However, the fundamentals have not changed significantly for a long time, so if you read this in later decades, they will be just as valid as they are now.

> The fundamentals of developing products have not changed for a long time.

1.4 The origin of products

Products have been developed and made for millennia. Originally, a craftsman would turn their idea into reality using the tools of their time. Some trained others how to do it. It became the basis of guilds and craft industries, for products crafted in small numbers.

The industrial revolution started the road to quantity production and de-skilling. Dividing tasks into parts, the division of labour, and mechanisation brought about a dramatic rise in the rate of production and prosperity. Technologies were developed and, with analysis of the production process, created the basis of today's manufacturing industries.

But in the West we began to lose our way early in the twentieth century. Large-scale manufacturers confused some of the principles that for example Gilbreth and Taylor had set out in analysing production processes and operations. We began to create unnecessary tasks and to introduce tortuous bureaucratic practices that added unnecessary cost to the product. Then came a ray of light in the early 1950s. In Japan, at first independent of each other, Ohno and Shingo started to reconsider the roots of the methodology that had originated in Western manufacturing. They examined the root principles of how it should be operated and the result was products that cost less to produce and attained high levels of quality. Stock and work in progress became labelled as evils; value stream mapping showed the way to single piece flow, and production cells rather than lines, made possible by drastic set-up time reduction.

Two seminal works summarise the methodologies and the outcomes. Shingo[1] describes how manufacturing process and operations can be analysed and designed to produce fault-free goods faster and at low cost. Womack and Jones[2] describe how the world's most competitive manufacturers have applied methodologies

[1] *Non Stock Production*, Shigeo Shingo.
[2] *Lean Thinking*, Womack and Jones.

based on Shingo's work (and other leading practice) to achieve immense reductions in cost and lead times. Both works are concerned with how the product is manufactured, not how products are developed. Many subsequent books have elaborated on this but these two explain the fundamentals most clearly.

1.5 Developing better products faster, at less cost

But manufacture – converting raw material into products – is only the final stage of the three key internal management tasks that face every manufacturer. The first two are developing products that customers want; and receiving and processing orders so the product can be exchanged for customers' money.

The three key internal management tasks are

1. developing the product
2. processing orders for the product
3. manufacturing the product

Just as the purpose of an animal is to catch dinner and procreate, developing a product is the purpose that started manufacturing. The origin was to commercialise the products that were developed. And order processing was necessary to complete the company's internal function.

In the same way that they have revolutionised manufacturing, some Far Eastern companies have taken strides in the way they develop their products, followed by a few in the West. But there has not yet been a concise explanation of the totality of what they do and why they do it, in the way that Shingo has set out his work on manufacturing processes and operations. Many works have described how leading companies accomplish parts of the product development process. But there does not appear to be a single volume that sets out a manager's guide concisely and which encompasses all aspects of these newer development methodologies.

1.5.1 The need for a management guide

Having spent 30 years in manufacturing, largely concerned with developing products, I would have valued such a guide. I often wanted to find fast guidance on the particular challenge that faced me at a particular moment. At the time, I often failed to find what I needed to apply to the situation that faced me.

I soon found that fellow managers in other companies also wanted a concise all-encompassing guide. So for my own use I started to create one. I started with my own experience, and systematically gleaned information from all the situations and works that I came across. I went through hundreds of references. I also added lessons learnt by others: with colleagues, I investigated how more than 70 really competent product developers accomplish every task. I continue to pursue that analysis, adding several companies every year.

This book is the result of that work and sets out the lessons learned. I also refer to works that competently set out key areas of background. The aim is to create an everyday help that managers can refer to at the various stages of the development process.

In due course perhaps others will be motivated to refine this management guide and produce an even more concise and easily referenced all-encompassing handbook for managers.

1.5.2 The purpose of this book

The objective is therefore a guide on how to set up and manage the process to develop better, higher quality products faster, with fewer resources.

As managers, we invariably have to focus on tackling problems of the here and now; it's often difficult to find time to research and sift through all the books and papers that could help, even if you carefully index what you've read in the past. I therefore set out the book in the order that you need to do the tasks. And I refer to sources of background detail if you need them.

The product development process:

Why you do it (*Chapters 1–3*)

↓

How you organise to do it (*Chapters 4–6*)

↓

The sequence of events (process map – *Chapters 7–9*)

↓

How you control it (*Chapter 10*)

↓

The tools and techniques that help you (*Chapter 11*)

↓

Source references (*Chapter 12*)

By absorbing the outline, you can amend your strategy piecemeal or totally depending on what you find appropriate. And having read the whole, you can refer back to a particular section that interests you now. But whatever you end up doing, you will need to follow the framework of tasks and use the tools set out here if you want to minimise risk and maximise profit.

1.5.3 You need an effective route map and effective tools

The background detailed principles of how you design products have been set out at length many times. Authors such as *Clausing, Pahl and Beitz,* and *Pugh* laid down sound principles and detail examples; authors such as *Reinertsen* set out excellent descriptions of certain tools and methodologies. I refer to them in the text where the user may want background detail. You will find a small library of these references in *Chapter 12*.

But detail on its own is not enough when you're faced with a real-life challenge. You usually want to find where you've taken or are about to take the wrong road, not analyse the road surface. You need a good map with some help to apply it.

You need to step back and understand the principle of what you are trying to do – and especially why you need to do it. And when you've understood, you need some guidance on sub-principles and sub-sub-principles. It's the overall plan and the underlying detail that distinguish the superb from the average. If you have your head down concentrating only on the detail, you may miss the signpost to your destination. Similarly, if you keep your head up and see only the main route, you will find that the detail prevents your getting there. You have to do both. You need to map out your main route first, with the main landmarks. Then understand what to do at each junction; then add the detail. And you need to improve your driving technique continually throughout the journey. In that respect, it's a journey you will never complete; you will ever be finding more effective and faster means to arrive.

The key is the process itself and the toolkit you use: the sequence of operations and the logic of why you do each step and what will happen if you don't. Some books set this out fairly simplistically. I found that many companies do it even more simplistically, frequently omit vital stages and do not make use of vital tools. Equally important is how you organise yourself to run the whole programme. You can't do it by punting tasks from one group to another (like most companies still do).

1.5.4 Who needs it?

If your company is small, you may wonder how this can possibly be relevant to you. Well beware. If you develop your own products that are successful, you are just the kind of company that will grow to medium or large size. If you become as competent at developing products as our exemplars, your growth is inevitable. Only companies like yours can create the wealth necessary to fund such growth. This is a worldwide feature, as you will see when you dig deeper.

The second reason it's relevant to product developers of all sizes and sectors is that you all experience the generic challenges described on these pages. The tasks you have to accomplish to achieve a successful product are related, whether developing a new chemical molecule or a new electronics or mechanical device. Although most of the examples are from the mechanical–electrical–electronics sectors, they read across to other sectors and to all company sizes.

For example, maybe everyone knows what everyone else is doing in a small company. (Only maybe: I've experienced small companies with higher Chinese Walls than big firms.) But start to grow and see what happens. 'Interesting' new tasks creep in like topsy to combat the communication 'challenges' invented by the increasing head count. Tortuous systems and procedures appear that increase the cost of production without adding any value to what the customer receives. And because it becomes increasingly inconvenient, and often the end objective becomes shrouded in fog, we all start to avoid or pervert the systems. Dilbert wasn't created out of thin air – real situations underlie the humour.

For example, many years ago when I worked for the newly acquired UK arm of a large multinational (Chrysler), a peer manager from the US Mother Company lunched with us. 'How do you ever manage to get the equipment you need using your tortuous procedures?' we asked. 'Simple.' he replied 'We steal it from another department then sort out the admin if they find out. Otherwise we'd always be late.'

> Leadership is the art of trading imaginary things in the future for real things today. –Dilbert

1.5.5 So where do you want to be?

Before you even start, you need to decide where you want to be. Most of us know the broad principles of what we should be aiming at. This book sets out the logic of why it should be done in a particular order, the effort and organisation you need at each step.

But before we examine the process, it's instructive to see why product development is important to manufacturing at all. Couldn't we just make what someone else has developed? Or just sell someone else's product to the customer? Developing products raises uncomfortable dilemmas. To many managers, the process of creating them appears costly and risky – an apparently bottomless pit threatening to swallow the company whole. Wouldn't it be easier to make a return on your assets doing something else? Some have, some failed; but in the end United Kingdom without product development could become destitute; not where you might want to live. Many have made serious money by developing and commercially exploiting really good products. And it's really satisfying.

Chapter 2
Wealth creation in the economy
Product development influences wealth creation

Why and how do firms that develop products have such a driving effect on the economy? What is the significance of intellectual property rights? What is the difference in economic effect between spending on research as opposed to development?

2.1 The origins of productivity performance

Most manufacturers just make parts of what someone else has developed. Although that in itself is a challenging task, it doesn't have the potential to create as much wealth as developing the product. Indeed, some companies just develop and sell the product, delegating manufacture to another company. Invariably they generate more wealth than the company that manufactures it. We examine why.

> How you develop your products is the root determinant of your profitability.

Developing a product generates intellectual property rights (IPR) for you, the developer. This makes a disproportionally higher return than just making what someone else has developed, where you don't own the IPR. As a result, you create higher added value[1] per employee (productivity). How high this level of productivity is depends, consciously or not, on how the product is conceived, designed and developed. Because that is where you plan the added value creation – before production starts. How you develop your products is the root determinant of your profitability.

2.2 How PDMs create economic growth

Figure 2.1 shows schematically how the economy depends on product-developing manufacturers (PDMs). Product development drives the economy. If a country has too few manufacturers that develop products, there won't be enough work to

[1] Added value is total sales minus total purchases; added value per head is the definition of productivity.

Figure 2.1 Product developing manufacturers feed work into the whole economy

support a large number of sub-contract manufacturers, which in turn restricts the growth of the service sector.

Manufacturers that don't develop products depend on those that do for their work. And half the service sector depends on the whole of the manufacturing sector for its work. The other half of the service sector works for the whole economy – manufacturing, service and public sector. This other half of the service sector would disappear if the manufacturing sector disappeared because its customers would have no money to buy its services. The service sector's prosperity thereby depends on having viable manufacturers, which in turn depends upon having viable PDMs.

The growth of the entire economy therefore hangs upon the fortunes of a small nucleus of PDMs to keep the rest of the economy profitably occupied; it's a fragile situation. But it's the key to driving the whole economy. So, aside from selling indigenous minerals or agricultural output, a country with few PDMs will have difficulty in growing its economy.

2.3 Interdependency and what initiates growth

Many politicians and economists argue that this is too simplistic a picture because '... all the elements of an economy are interdependent.' And furthermore, they argue, if all United Kingdom product developers were to disappear, our manufacturing sub-contractors and service industry would still thrive because they would just serve other countries' PDMs.

The PDM, product-developing manufacturer, drives the creation of economic growth.

Is this realistic? The United Kingdom needs export revenues to buy imports because we are not self-sufficient. So, could our service industry and manufacturing sub-contractors provide all the revenue we need by exporting their services? It is widely acknowledged that they could not.

First, over 95% of international trade is in manufactured products, not services. And second, location is key: the town corner shop would be challenged to exist on customers from the next town, no matter how near. In the same way, other countries don't need much if any of our sub-contract manufacture and service industries' output – they have their own.

It was international trade that caused economies to grow. In the beginning, international trade was entirely in goods and there was virtually no service industry. At first we traded indigenous material, agricultural products then, later, manufactures. More than 95% of world trade is still in manufactured goods; services feature little, but are slowly increasing.

The economist's argument that the elements of an economy depend on each other is true. An example of interdependency is the human body: all its separate elements are inter-dependent. But some are more important for survival than others. It's a challenge to survive without heart or brain (no wisecracks about politicians, please). In just the same way, PDMs equate to the heart or the brain; survival is a challenge without them.

Only a small proportion of manufacturers develop the products they make. For example, although in 2003 there were around 140,000 manufacturing entities in the United Kingdom, we researched that products were developed probably by fewer than 15,000 – not much more than one in nine. No one knows the exact number because the data is not collected. You may think that having so few PDMs is unique to your area. It isn't: it's widespread.

Small companies that grow rapidly almost always start life by having an idea for their own product, which is how they fund their growth. Those that don't mostly stagnate as small sub-contractors. In the United Kingdom, for example, only a small percentage of sub-contract manufacturers employ more than 30 because their return on capital is so small that, with few exceptions, they can't fund their growth. Whereas most PDMs grow to more than 30.

Successful PDMs invariably grow and create employment. It is highly wealth-effective when they grow because of the amount of work they cascade into the whole economy. If they grow, the whole economy grows ten-fold their growth or more.

2.4 Why not invest more in product development?

Manufacturing industry has to be price-competitive with internationally traded manufactures. But service industry generally does not, because services are mainly not internationally traded. So service industry can set its pricing within a region without pressure from imported services – it can operate as a kind of local, informal complex monopoly. This is how the UK service industry averages to return more than 20% on its assets. For example, UK hotels charge far more than those in France despite lower UK labour costs; some UK lawyers are the most expensive in the world.

Many UK service companies routinely increase their prices by twice the rate of inflation (sometimes more). Whereas most manufacturers' pricing is constrained by the price of foreign competitors' goods. And many manufacturers are expected to reduce their prices by 2–3% every year through continuously learning how to produce more cost-effectively. So UK manufacturing industry can't operate as an informal local monopoly. As a result it averages less than 9% return on its assets.

A further difference is that rapid rises in currency strength can cripple manufacturing in the short term because it is traded across borders, and a stronger currency immediately makes it less price-competitive. In time, it can adjust its processes to respond, but not instantly. Whereas a rapidly strengthening currency has little immediate effect on service industry.

Many investors believe they will make a better return by investing in service industry. (In reality, product developers tend to return nearer 20% compared to most sub-contractors' 5% or less.) This deters investment in UK manufacturing, and especially in R&D within manufacturing. So in effect, the service sector tends to starve the manufacturing sector of investment funds.

Investment in manufacturing is essential for long-term solvency.

But investment in manufacturing is essential for the long-term solvency of UK plc, as well as for the economic growth of any country not dependent on mineral or agricultural wealth. Why? And how can you redress the un-level playing field with service industry?

2.5 What initiates economic activity: how do you feed it?

As we argue above, it's not enough to believe that all sectors of the economy are inter-dependent. You have to identify and feed the mechanism that initiates activity: to increase gross domestic product (GDP) and employment, you need to encourage the activities that *initiate* growth. PDMs are the key.

The only way PDMs will grow is by developing better products that win more sales. If they can perform better, and if non-product-developers can achieve the motivation to start developing products, you create sound, sustainable, economic growth.

More than 80% of new work that enters the UK economy is initiated by a small number of PDMs, not by sub-contractors or service industry, who depend on manufacturing for their work. Only a small amount comes from service sector exports such as consultancy and the financial sector. And although the public sector can commission work, its effects tend to be temporary and can divert resource from manufacturing investment. Ultimately, too much public sector investment can affect international competitiveness, taking years to work out of the system.

2.5.1 How you feed the growth mechanism

To induce PDMs to justify spending more on product development, which increases their sales, some countries use product development (or R&D) tax credits. It has been found that a tax credit of x% results in x% more product development activity. This increases the number of PDMs that grow into larger ones because in general only PDMs can generate the added value to fund growth from small to larger. That is why you don't find many large sub-contractors. It's also highly cost-effective for the country's treasury because it disproportionately increases their GDP and therefore tax revenues.

In the United States, for example, for decades product development tax credits have encouraged their companies to develop products. This has helped them to fund growth that has fanned out through the entire US economy. The tax credits have had (and continue to exert) a strong influence on the growth of their manufacturers. This, and the more ready availability of venture capital, could help explain why, relative to the United Kingdom, the United States now has three times the percentage of 100–500-employee companies. These medium-sized companies are a powerful growth engine for the whole economy.

Product development tax credits are also useful to help redress the un-level playing field between service and manufacturing. Why? Because manufacturers can justify proceeding with product development projects that have a lower potential rate of return (IRR).

Companies often have a range of possible projects offering different IRRs. The higher the IRR threshold, the fewer that are approved; the fewer that are approved, the lower the company's rate of growth. A tax credit has the effect of lowering the IRR threshold at which projects can proceed; more projects meet the IRR target threshold, which thereby stimulates growth.

Arguably, these credits should not be allowed against research that is not related to products, or on design work that is corporate identity, annual reports, advertising or the like. Many manufacturers spend more on these than on developing their products, with predictable consequences: well-promoted unattractive products rarely dupe consumers.

2.6 The financial effects of spending on *Development* rather than on *Research*

'D' is the design and development of technologies and features of the next product. The United Kingdom and United States spend only around 30–40% of their R&D expenditure on D – a small proportion compared to Japan, which spends nearer 70%.

> The financial effects of spending on D are fundamentally different from spending on R.

'R' is research – longer-term research into blue-sky science at one extreme, for example, in universities. The United Kingdom is good at this; but bad at getting a financial return from it. The United Kingdom incurs a huge net financial deficit on funding science research. Whereas company-funded applied research into background building blocks for products maybe 5 years ahead, gives a good financial return. But this is product-related and should really be included in product development, D.

A lot of UK fundamental research is used by foreign companies that develop it for use in their products, whereas few UK manufacturers do a well-documented phenomenon. The United Kingdom thereby funds many foreign companies' R. This is an interesting factor to reflect on. It's often less expensive to look around for and to buy exploitable research results than to do it yourself. The European Union has a vast bank of such work that can be accessed through the European Union website (www.cordis.lu). Look for the results of past projects taken to demonstrator stage that are seeking collaborators to exploit them commercially.

And if you encourage D, you create spin-off growth throughout the whole economy. It also helps fund companies' R. If you encourage R, you don't create growth until the results are incorporated into a product. They may never be. You therefore need to give incentive only for D.

PDMs are the ones that perform the D activity. A few also perform some product-related R. It is they that *initiate* growth in the economy; no other sub-sector has such a great capability to do so. How countries create wealth has interesting lessons for individual companies, as we explore next.

Chapter 3
Your company's fortunes
How product development affects your company's fortunes

How does product development create wealth for the firms that do, as opposed to those that don't? Why is it not enough just to manufacture – indeed, do you need to manufacture at all to prosper? And how do you develop your capability? What makes a difference and what doesn't?

3.1 Profit and product development

As Ohno (see references, *Chapter 12*) points out, 'Your customers neither know nor care what it costs you to manufacture what they buy'. They just assess their purchase decisions on perceived value.

The market decides what it will pay for your product. And you control what it costs to make. The difference between the two is your profit. And since you can't extract more than the market is prepared to pay, the only way to continually improve your profits is to reduce your costs continually. Continual cost reduction is therefore key to survival and growth.

It is clear however that many companies don't understand this. And equally, they don't understand that it is the product development process that initiates most of their costs in the first place.

3.2 Product development: the way to generate wealth

Considering that you won't make real money unless you develop your own product, it's surprising how few manufacturers do. In the United Kingdom only about one in nine does; in the United States, maybe one in seven. Why so few?

> You won't make real money unless you develop your own product; too few manufacturers do.

A common belief is that acquiring other companies is a safer, quicker, cheaper way to grow. But 70% of mergers and acquisitions are a financial failure. Many adopt the wrong strategy, buy the wrong companies at the wrong price, integrate them poorly and fail to reap synergies that were often imaginary. It can result in far more business effort than becoming a leading product developer. And it often reduces the capacity to innovate because hugeness of scale often stifles creativity. But maybe it looks easier because it's taught at business school whereas how to develop products is hardly taught anywhere.

> Pursuing acquisitions rather than learning to develop products well is not the quick fix it appears to be.

Some think that developing products is too risky. Some burnt their fingers badly and are afraid to do it again. Others think it costs too much so they put it off. The feeling that new products take a long time to return their investment adds to the higher apparent risk. These beliefs, among shareholders as well as company executives, can make the idea of new product development seem like a black hole that could swallow the company. But these beliefs indicate an unawareness of what is really possible. Developing products has the greatest potential to generate wealth. We all need to learn how, not avoid it. Companies should focus on what they do best and do it better than anyone else.

> Nearly everyone spends up to half their product development budget on activities that add no value whatever to what the customer buys.

But even among those companies that do develop products, most could do it so much better. For example, nearly everyone wastes up to half their product development budget on activities that add no value whatever to what the customer ultimately buys. You may not realise you're doing it. Not only that, but many firms run their product development such that customers won't like their product – it guarantees they'll buy someone else's. You're bound to know products you wouldn't buy yourself.

You may think your problems are unique. They're not: all industry sectors experience the same generic problems. For example, most have great difficulty in defining what customer wants. Few have a really good route map for their development process. Consequently, most over-run time and budget, often through poor project management. Even fewer understand how to use the tools that could help them cut time and cost.

And yet developing new products is a really cost-effective route to growth. Those that do it well know how to manage and control the risks head-on. They carefully develop a strategy for new products and gain more advantage by making use of their inherent strengths.

So what can you and we do about it? How can we become as good as the best, to use our budgets to best effect and launch products that make good sales and real money?

3.3 The effect on sales and growth

What you develop is your future: your company's business strategy needs to be product-based for survival and growth. Your long-term profitability depends on continually improving your technology capability and competence; and your ability to exploit it in your new products.

> The most profitable companies' success comes from their superior product development capability.

Despite its fundamental importance, few manufacturers in the West have product development capability. And yet the most successful newly industrialised countries are building their economies on it. Their products are taking business from the West both in home and world markets.

But product development effort is effective only if it is part of your company's long-term strategy. Otherwise it risks being under-resourced, under-supported and financially ineffectual. The most successful manufacturers have their product development programmes enthusiastically championed and supported at board level.

Probably fewer than 1% of Western manufacturers are world class. Interestingly, most of these believe they still have a lot to learn. Of the rest, about half have the potential to be, if only they would take the necessary action. But most lack motivation; although they aren't satisfied with how they do it, they act only when failure faces them. The less said about the final 49%, the better; most of these reckon they're good at it!

> The highest profit growth companies' advantage comes from their superior product development capability
>
> - With a more focused strategy and a predictable process, they get to market sooner through better control.
> - They develop products that generate better margins because they prioritise effectively and therefore have less in the development pipeline; but they spend far more on each project.
> - They have far fewer change orders and manage them better; with even fewer changes after production release.
> - They know how to generate innovation without compromising launch dates.

There is a major common factor in the most successful manufacturers: their product development programmes drive their success. They develop successful products not just occasionally, but as part of a continuous programme. They have certain characteristics:

- their chief executive champions their product development and makes sure all his fellow directors understand and participate;

- they have an open, no-blame, internal culture that encourages analysing and learning from mistakes;
- they understand how to discover their customers' innermost needs;
- they generate excellent product ideas, formally and informally;
- they continually develop their technology and product development capability to produce superb products from these ideas;
- they have an excellent route map and programme management.

The top 2% introduce a continuous stream of new products and have few products that are older than three years. The most profitable 10% have 70% of their product portfolio less than five years old. Typically, the fastest growing companies have such attractive products that they double their sales every three years and export more than 25% of sales.

In the West, since the start of the 1990s, some industry sectors have been pressuring their suppliers to become product developers. For example carmakers are telling their suppliers to acquire product development capabilities if they don't have them; and to improve them if they do. The suppliers' future business will depend on their designing and developing the component systems and products their customers need – or they won't continue to be suppliers.

3.4 Just manufacturing is not enough

Unless you develop your own product, you have little or no intellectual property rights that you can charge a premium for. Sub-contractors usually make far lower margins than product developers – I discovered this the hard way from running companies that did both. First, sub-contractors can't create as much added value unless they own a unique process; all they can sell is their man and machine-hours, plus their level of service. And second they are subject to fierce competition from lower cost competitors, sometimes from developing countries that have far lower labour and machine-hour costs. It's difficult to compete against companies that have labour costs that are only 5% of yours. You need an extra attraction – your product's benefits and an emphasis on more up-market products that have better margins and are less pressured by Far Eastern competitors. (Discussed further in section 3.9.2.)

Having said that, a lot of product developers make poor margins because they could do it so much better. Some try to claw back margin by having parts made more cheaply abroad. Their poor product development usually starts with not researching what their customers really value. And most have an inadequate process map that leads to a poorly project-managed process, as well as to poor teamwork, and often do not use tools that could cut time, cost and risk. They spend too much time on tasks that add no value, are usually late and, from inadequate input to their design process, their products costs too much to make. And too few make effective use of the design talent that is available within the West; many Far Eastern companies use UK talent better.

3.5 Its strategic importance

If you rely on yesterday's products and technologies, competitors you'd never heard of a few years ago will take your sales. Product development is necessary not only for growth but also for survival. The most vulnerable kind of company is the old-established one, with massive investment in manufacturing capacity and a departmental structure. They once led in a technology that has now been stretched to the limit of its application. They may make good short-term returns because of their market share and scale of manufacture. But in reality their products are incremental developments of old technology. If this sounds like your company, beware. Industry leaders can no longer feel secure just from their current technological knowledge. They must continually develop new skills.

> With the increasing rate of development of technologies, in the end no company is safe with its traditional product portfolio; and technology development is the least sustainable advantage because of worldwide communication; so you have to pursue it relentlessly.

As products mature, more countries learn how to make them. They become 'commodity products' such as white goods and motorcars. Not a good sector to be in: such products are very cost sensitive. Far better to continually upgrade your ideas and technology to make sure you are at the top of the pile, continually making new technology, innovative products. Those that don't soon find that they face unwelcome financial challenges.

3.6 Product development sets your potential productivity

To achieve high productivity (added value per person) requires you to create high added value (sales less total purchases) in the first place. This implies products that are more up-market. Around 90% of a company's added value performance is planned during the early stages of the product development process. It is more difficult to make good profits to finance growth if you only manufacture parts to product developers' drawings.

> Product development is a company-wide process that determines more than 80% of your future business performance. It locks in over 80% of your products' costs, quality and performance. It governs your growth rate.

Common wisdom is that productivity originates in the production area, from an efficient, low-cost, lean and well-trained production facility. A lean[1] organisation is

[1] Maybe fewer than 100 firms in the West operate true zero-stock production.

essential. But in the first place your product must be cleverly designed to maximise added value and minimise cost. So if your product development process is ineffectual, you won't create as much added value. Your potential productivity will be lower.

As the schematic shows that although spending in the early stages is low, the decisions made then commit expenditure later in the process. For example, although production plant may not be resourced until later, the product's concept and subsequent detail will determine the cost of that plant. So inadequate brainpower applied to the early stages can cause high levels of cost later, which could have been reduced by better early design stages.

A study by Hull University found that 80% of the quality of a product was originated during its development. They measured the level of quality by the number of defects reported in production and in service. They also commented that it cost up to 1,000 times as much to put them right as to have resourced a little more input at earlier stages to avoid them.

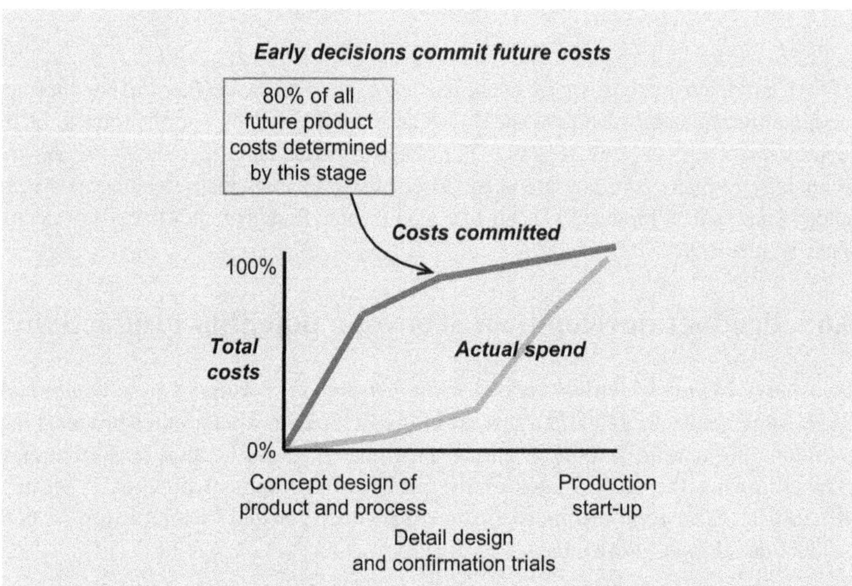

After a product has been developed and the manufacturing instructions issued, the manufacturing area by itself cannot originate much more added value (and hence productivity) without making changes to the product (although it can add a lot of costs through wasteful tasks.) But having to make a large number of changes incurs high costs and also indicates that something is wrong in the way you organised your product development process (examined in detail in 9.4 to 9.7). And the later your changes occur, exponentially the more it costs you.

3.7 What makes for effective product development

Your company's survival and prosperity may well come to depend on your having an effective product development capability. It is key to

- producing the level of features (and technology) your customers want;
- achieving low production costs – lean products for lean production;
- minimising production support overhead (unnecessarily complex products need unnecessarily costly control);
- developing and producing them on time.

The most successful managers involve all company functions in every aspect, from investigating what customers want to launching the product. The best plough back heavily into their programmes and recruit and nurture the best people at all levels.

> Driving product development is the most important contribution a chief executive can make to their company's success. – Bill Mayer, VP Philips SDA Division

The factors that create the product's success are

- customer focus (from end user to your immediate customer);
- better technological competence than your competitors;
- an effectively managed structured process using a cross-function team;
- good understanding and use of tools and techniques;
- good information management.

Poor capability spells poor sales, poor profits and a poor return on assets.

3.8 Give customers what they value

A product is not a stand-alone artefact. It is a means of solving customers' problems. In some sectors, three quarters of all successful new product ideas come from customers.

Successful projects focus from start to finish on what the customer will buy. This is invariably underpinned by the company continuously developing the product's building blocks. We emphasise the need for a low-cost continuous background development programme below.

> The only reason customers buy your products is because they see them as the best value means of solving their problem.

The only reason customers buy your products is because they see them as the best value means of solving a problem they have. Their problem may be real or imagined, physical or psychological. No problem means no need for a product and therefore no sale. They need to recognise that your product will give them the best solution, with unique benefits and perceived value, better than other products to make them buy yours.

But customers' aspirations increase continuously. Your competitors find ways to provide ever more sophisticated solutions at less cost. So you need to develop more desirable, less costly products continuously.

But most manufacturers (95% in the UK) don't investigate what their customers need or want at all. Many just look at what their competitors are doing; others develop what they think customers would like without checking. Both have led to interesting failures – most are lost to history but some are remembered such as the Sinclair C5 car or the Ford Edsel.

You not only research your potential customers, but you also develop products that exceed their expectations. Most customers are part of a chain, from raw materials' producer, through sub-contractor, systems developer, original equipment maker and buyer to final user. You may be somewhere in the middle of that chain. You need to research every level above you, not just what your immediate customer (the one above you) wants.

The customer chain:

- product user
- product buyer
- product assembler
- major systems maker
- sub-systems maker
- piece-part maker

Markets can become complacent and behave like commodity markets with little differentiation between products. Here, all the companies spend around the same percentage of sales on product development, often only half to 1% of sales revenue. History shows that if one company decides to spend significantly more, perhaps between 4% and 6%, they upset the settled market by suddenly taking more market share.

But because of the surprise element, competitors often take 2–3 years to respond. And once they lose their market share, it can cost them ten times as much to regain it as to have taken prior action to prevent its loss. So the new leader often retains and consolidates their lead and profitability despite the others' counter-measures.

The UK domestic vacuum cleaner market in the '90s was competing largely on price, all products being perceived as offering more or less the same. Then along came Dyson with a well-styled cleaner that didn't need replacement dust bags; it centrifuges the dirt into a transparent emptiable chamber – you saw when it was full. This took the somewhat sleepy market by surprise. In a short time it displaced the market leader, who took more than five years to develop an effective response (then lost a patent action brought against them by Dyson). Products are solutions to customers' problems, not just artefacts. Dyson reckoned that customers found buying and replacing the dust bag a chore they could do without. He offered them a solution with unique benefits that they valued. And was able to charge a premium price.

3.8.1 Generating fresh ideas

You need a mechanism to foresee change – to sense market evolution, to monitor customers' views and tastes and anticipate (do something about) technology developments. Companies continually disappear through not doing this. Some just hide their heads in the sand.

> Unless your new products have a 'newness' compared to your old ones, customers will switch to other companies' products that have.

You need to encourage your staff to monitor new developments in components, technologies and anything that could be applied to make your products more interesting to customers and easier to manufacture. You could be caught napping by changes and technology leaps that would affect the economic viability of your products. Monitor changes in life style if you are supplying direct to consumers. Tastes change; social consciences change; legislation forces change. You also need a background product development programme to evaluate what could be possible – see section 3.9.

A customer-driven product specification is the first key requirement. Don't confuse price with value. Customers don't want the cheapest basic product: more than 80% of our imports are from high technology, high wage–cost economies producing sophisticated, well-designed goods that offer a high level of features at an affordable price. Few are low cost, basic goods.

3.9 Developing the product

Product attributes are controlled by the design process – cost, features, performance, reliability, warranty costs, ease of servicing and more.

Your company's cost base is largely governed by the way the products are designed; the process controls not just labour and materials but your investment costs and many other direct and indirect overhead costs.

A hydraulic equipment manufacturer embarked on a major production investment. They had quickly designed and developed a new range of products that were very complex and demanding to manufacture. In fact they were unnecessarily complex. Not only was the investment large, but operating the new plant incurred a significant indirect overhead in the form of support personnel costs. It worked, but to survive they had to trim their profit margins. The financial trauma was such that the company took almost a decade to recover its profitability. Had they spent only 0.5% of the investment in planning the project and developing the design concept better, they could have produced a simpler new range with little change to their existing more conventional machinery. And many years on, they still don't understand what they did wrong – they could repeat their mistake. It's easy to design a complicated product; it takes far more effort to devise a simple, elegant low-cost one.

Your technology, innovation and development competence are key to minimising your costs.

3.9.1 How you develop your capability

Effective product development requires the best available brainpower. It's what determines the company's future cost base. Development covers three aspects:

1. Maintaining current products.
2. Developing the immediate next product.
3. A background building-block development programme that feeds de-bugged new ideas, features and technologies into products beyond that.

> You have to develop new product and technology building blocks continually. But do it offline until they are sufficiently risk-free for use in a product. Remember: people like to buy things that work.

Continually developing new building blocks is fundamental to the success of your future products – your future competitive advantage. A continuous background development programme is your source of tried and tested new features. Why background, offline? Because if you put untried ideas into your new product time-line, you wreak havoc with your programme timing. You cannot define precisely how long it will take to turn a new idea into a reliable, product and production-ready feature. That's why new ideas must be tried out and de-bugged offline until they are sufficiently risk-free to incorporate into your next new product. And new techniques, configurations and technology must also be investigated, understood and tried out – also offline. Use technology road mapping to find out what you should be doing and when (see section 7.2).

It's a means of managing the risk. And a background programme can cost you very little – maybe only 2% of your product development budget. If you do it well, you can continually leapfrog your opposition to arrive at market leadership far faster than by not formally managing the situation.

You can use your existing core competencies to attack new markets. But you can consciously work at extending your range of core competencies, not by acquisition but by developing your current capability. You use your background development programme as the vehicle.

3.9.2 Your strategy

Where does your company need to be in its level of product ideas and technology to achieve its business direction? You set out where you want to be. Your tactics will determine whether you develop all or part of your product in-house, contract it out, do a combination of both or buy licences. You assess the costs, time and risks of each.

You consider the technologies that the business is based on at present, that

- differentiate you from your competitors;
- are not yet ready for use but will give you competitive edge;
- are the support technologies you buy or use in the process.

You continually investigate and justify a series of elements

- to monitor how rapidly changes are occurring and whether there are discontinuities that need plugging;
- to ensure that the company's technology base is consistent with the direction of the company's strategy;
- that you are not missing opportunities that could be taking place outside your company's current experience;
- how much to invest in a particular new technology.

It's an important formal activity.

A multinational automotive supplier refused to believe there would not be a permanent demand for their valve-operating pushrods. European car makers turned to overhead camshafts; we could see it coming more than a decade earlier. The company could have started developing products the new engines would need – such as hydraulic tappets for overhead camshafts – there was little European capacity at that time and they had a decade to do it. But instead they chose to do nothing. They didn't want to consider upgrading their level of technology. In the end, they had to close down the entire facility; and the costs that resulted from closure turned out to be greater than pursuing the alternatives. But most significantly, they lost a large chunk of irreplaceable sales revenue.

3.9.3 Your tactics

Your tactics are to nurture and grow your in-house expertise continuously. It provides the reservoir of long-term intellectual property that fuels your added value performance. If you don't, shallow knowledge will lead to shallow products that customers recognise surprisingly quickly. Better products from competitors with a better understanding will out-gun you.

To create a new product, you need to add or replace only a single new building block: incremental progress is fast, inexpensive and relatively risk-free, not a high-risk step change. See section 11.4.

The expertise you need has two aspects:

1. background development to seek and prove out ideas and technology for future products (maybe 5 years ahead); and
2. design and development of your immediate next products.

This is loosely referred to as Research & Development. In reality it's two separate activities: research and technology, and design and development.

You need continuous effort to underpin your products' development by working on the technologies these depend on. But you don't need to do all (or any) of your own research; some may cost you nothing more than your time to investigate what others have done; some can be bought quite inexpensively. The EU Cordis website (www.cordis.lu) lists many.

Finding, assessing and applying information is often a challenge for smaller companies. The more resourceful ones overcome this by for example working with larger non-competing companies. Japanese companies are particularly good at doing this.

A small firm developed new infrared paint curing equipment for the automotive re-finishing market. Among other factors, it used a carefully considered technology policy, which led to doubled sales in three years. A key feature of the new equipment was making use of a major multinational's technology power in developing unique radiant heaters. It helped this small company to present itself and its products as best in the world. It achieved 90% of its sales income from exporting its products. It could not possibly have done this by funding the technology development itself.

Other companies contract research work to universities and research organisations. And often, Government bodies will help.

In the early 1990s, Brook Hansen Motors, in response to market pressures, needed to develop electric motors that were more efficient but cost no more to manufacture. Analysis indicated that it was theoretically possible to improve efficiencies by about 3% without significant cost increase. But it required higher-performance low-cost electrical steel combined with a radical overhaul of motor design and manufacturing methods.

Their technical director, Professor David Walters, realised that strong project management on its own would not be enough. It would need additional research and development input to meet the time scales.

Orb Electrical Steels agreed that it might be possible to develop a steel having the required magnetic properties at or near the target price. Both companies considered that the commercial risks and the R&D costs were too high, so they successfully applied for grant aid under the UK Energy Efficiency Best Practice Programme. In addition, fully funded research assistants were retained at Sheffield and Cambridge Universities to research electromagnetic design and iron losses. And a thermodynamics specialist from TEPRO, Prague, worked on air flow and fan design for heat dissipation.

The outcome was their higher efficiency 'W' range of motors, introduced on time and to their target costs.

3.9.4 Should you do research as well as development?

The amount of technology background development or research you do will depend on your sector. Pharmaceuticals and aerospace will spend far more than mechanical engineering, for example, because of the relatively high new technology costs.

You don't necessarily need to spend large amounts on research; but don't ignore it.

The balance between your background and current development projects will depend to an extent not only on sector, but also on the maturity of the technologies you use, and whether you aspire to be a leader or a follower.

A leader in a mature sector might spend 5% on background and 95% on current development. Whatever the proportion, it should be enough to avoid your customers thinking that you're falling behind. A company with a programme to overtake the current leader may initially fund more background work to catch up.

Each new generation of technology tends to cost about 15% more than the previous one. This causes added value per person and profitability per unit investment to increase. There appears to be no internal limit to your becoming more competitive by developing your technology and knowledge. Where you sit in the total spectrum of profitability depends only on your internal efforts to exploit this means of growth.

3.9.5 Protecting intellectual property

As part of your long-term product development programme, you may generate ideas that you want to prevent others from commercially exploiting until you have milked them. There are many ways of doing this apart from the high costs of using patents.

Patents are cost-effective for ideas of fundamental principle, for example a new principle of operation of an electric motor that needs no windings. But they are less useful for protecting detail, for example, a new way of configuring the windings of an electric motor. Detail patents start a defensive patenting war with competitors who find small deviations from your detail patent that also work. Only the patent lawyers win. Consider alternative forms of protection, such as design copyright.

A useful website on patents information is into the European Patents Office in Vienna via www.european-patent-office.org or through www.cordis.lu/ipr-helpdesk. You can search for the technical subject (field), text in the title and so on. It's often less costly to buy the rights to use an existing patent. The world is full of unused ideas. And it's often easier to circumvent patents by taking their idea and modifying the detail. The Japanese do this especially well.

If you have a brilliant idea and want to protect it, it can cost tens of thousands of dollars to achieve comprehensive worldwide protection, plus periodic renewal fees. At the other extreme, you can do it yourself for a few hundred dollars to get limited single country protection.

But consider your long-term strategy; will patenting generate an acceptable return for your outlay? Is it strategically necessary? Or more importantly, can you afford to take infringers to court? Unless you have something really fundamental, it's often more profitable to develop products so fast that you're always ahead of your competitors.

3.10 The structured process

Just having a formal new product development process, on its own, has no commercial impact. What is important is the detail, and whether people operate it fully – you have to have the right culture, the 'soft' issues. It's the culture that determines how well you operate, not the procedure manual.

The leaders cut time and cost by putting far more resources into the upstream processes.

The leaders' emphasis is on pre-development work. They spend more time and money on the upstream processes and far less on the downstream. They start with a quick market and technical assessment followed by a more detailed business case investigation. They achieve an early clearly defined marketing specification and avoid delays from vague, moving specifications. They accurately define the investment and capabilities required, and control the programme to hit their time and cost targets. From thoroughly researching customers' needs and aspirations, and using state-of-the-art development abilities, they produce products that exceed their sales targets, produce a fast return and increase market share. We examine how this is done in detail.

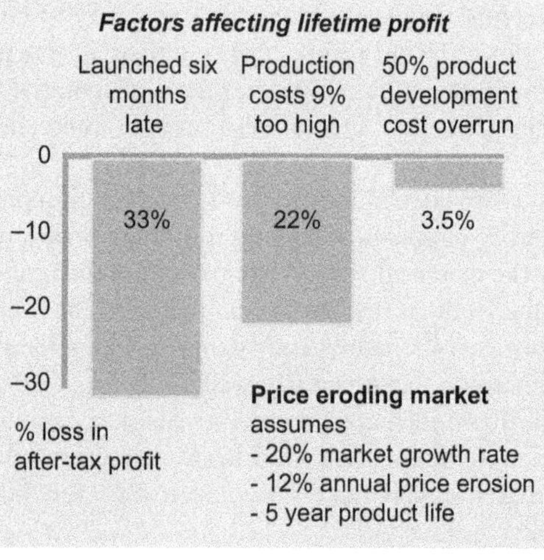

A well-structured development process is essential or you won't be able to develop products to programme time and budget. Lateness usually kills profitability even more than a design that costs too much to manufacture. Spending money to achieve your time and cost targets costs less and has only a small effect on the product's lifetime profitability. This is examined later in section 10.4.

Companies need to use the tools and techniques that can help them to manage and get the most out of the process. Few companies have a good understanding: again, dealt with in *Chapter 11*.

But above all, you need the support and interest of the CEO to set the culture and ensure that the structured process is operated at all levels. Without this it won't be, which unfortunately is the norm in most companies.

3.11 How much should you spend?

Many manufacturers spend more on their annual report and product brochures than they do on developing their new products – they are not only bad at managing costs but even worse at allocating resources.

> There is no typical spend figure. Most important is not your absolute spend but how much compared to your competitors, and how effective.

It depends on the sector and the maturity of the technologies. In a mature market, such as white goods, companies may spend around 1% of sales revenue on product development. A company wanting to shake up the sector up may spend 5–10% on a novel new product range and force the competition to respond. In less stagnant markets, such as the machinery sector, a 5–10% spend is not unusual. At the other extreme, in the newest technology sectors, some companies spend 50%. But that figure will reduce as the market matures.

In the 1980s, the European iron foundry industry was typically spending 0.1% of sales on developing new products or processes. One continental foundry equipment maker reviewed its policy and decided that this was such a derisorily small amount that it would allocate 1% – still a very small figure but a huge percentage increase. The resulting advances in its products revolutionised the application of specialist cast irons in the automotive sector over the next decade and rendered many competitors obsolete.

It's not just the amount but the effectiveness of your spend which is all important. Most companies incur a large proportion of the cost on tasks that don't add value for the customer but which they charge them for. Your customer neither knows nor cares what it costs you to make. What the customer seeks is value. You could be vulnerable to competition from other companies that understand this and don't incur such high costs. They will either undercut you or make a far greater return on their assets than you can on yours.

You could do twice as much with your existing development cost base. You need to structure what you do to avoid wasting time on tasks that reduce your income. It applies to every manufacturer: don't think that there's no scope for you. See *Chapter 5*.

Most companies could achieve their current performance using only half their resources.

If your firm's production area has not adopted the principles set out by Shigeo Shingo to produce the benefits described in *Lean Thinking*,[2] they probably have the potential to halve their floor space and overhead cost. It's not about old-fashioned cost cutting or about business process re-engineering. A lean methodology reduces effort and cuts waste, doing twice as much with your current resource. Rother and Shook show how to approach this in *Learning to See*. It's a methodology that only a few hundred manufacturers have adopted worldwide – and they're cleaning up on market share, profitability and low cost of operation. For example, few need manufacturing resource-planning systems (MRPs) when they've introduced lean practices.

Many companies have also halved the time they take and the money they spend to develop their products, refined from frequent practice. The best now do it for a quarter the cost of a decade ago – and still decreasing.

The more often you do it, the faster and less costly the process becomes.

Project cost is very much time-related: the longer it takes, the more it costs. But the faster you do it, the less it costs and the more accurately you meet customers' needs. Your customer research will be more up-to-date with less time to drift. Speed and cost are paramount. If you take too long, technology and customers' needs overtake your specification. If a product takes too long to bring to production, you are getting no return on an investment that is increasing. It loses profit opportunity and absorbs more money funding its development.

You won't be able to reduce costs unless you understand where and when they are incurred. It is the product development process that pre-determines the bulk of your company's cost base – see section 3.6. As you develop a product, you progressively fix a higher and higher proportion of your company's future costs as more is pre-planned and less is left to chance. And as you move through the process you progressively decrease risk by eliminating the problems that lead to costly changes later. And the more you do it, the faster and cheaper it becomes through practice.

Your product development programme determines 80% of your products' costs, quality and performance.

[2] Shigeo Shingo, *Non-stock production*; and Womack and Jones, *Lean Thinking.*

Less profitable companies, by contrast, will often have fixed less than half their total costs to launch by the end of their design and development stages. They then add progressively more unplanned cost and time to their programme because of tasks they have unknowingly omitted in earlier stages. They don't have a complete process map to follow and they don't control what they do well enough. They over-run their budget, miss key dates and incur debilitatingly high warranty costs. Two-thirds of product developers are in this situation. And many don't have programmes to replace their current products when sales decline.

3.11.1 Downsizing

Downsizing is usually done because management fails to identify and cut waste continuously. It can downsize capability by cost cutting that ignores the potential to improve profit through creating growth. It often shrinks new product programmes. This usually damages the company's capability and cuts the rate of growth.

> Business Process Reengineering burgeoned during the recession of 1991–1992, when many big companies were desperate to cut costs as least as fast as sales were falling. The idea, popularised by Michael Hammer in **Re-engineering the Corporation**, was for special teams of process re-engineers to analyse key processes, identify the waste, and quickly remove it to create smooth flowing processes at lower cost.
>
> But most of the re-engineers (mostly outside consultants) lacked a credible method or any experience and gained little co-operation from employees. Few processes were successfully re-engineered. In the end, many employees were laid off to meet the consultants' promises to management for almost instant investment paybacks.
>
> The 'three Rs' often became Retreat, Re-organise and Retrench (or downsize, rightsize then capsize).

It can produce a dramatic, immediate but short-lived increase in the rate of return on assets. But profitability usually declines after a honeymoon period and progresses to significant profit reversal in the medium term. Some downsized companies have bombed down their sector performance league table. Some have experienced total failure in the longer term. See left for explanation.

To generate more added value, you need to improve capability not reduce it. Follow the lean thinking approach; cut your costs without reducing capability.

3.11.2 Private equity 'turnarounds'

Compare a traditional private equity 'turnaround' with a 'lean transformation'. In the former, the objective has been to produce a dramatic bottom-line result quickly. This has often meant

- rolling up companies in the same industry to reduce competition and increase prices to consumers;
- negotiating lower wages and benefits;
- cutting spending on long-term development projects not critical to the firm's strategic plan;
- reducing headcounts in activities judged non-essential;
- restructuring the balance sheet to add bank debt, often creating instant dividends for the private equity firm but high levels of long-term debt for the firm once it is sold;
- re-negotiating prices with suppliers, on threat of loss of business.

These actions move wealth from customers, employees, suppliers and former owners to the new owners. In some cases this may avoid a firm failing. But it is often unclear that any additional value has been created in the sense of better satisfying customer needs with a given amount of human effort and capital investment. And, from society's standpoint, the only way to increase living standards is to change the ratio of human effort and capital going into a firm to the amount of value coming out. Otherwise the outcome is a zero-sum, with some winners and some losers.

The only way to increase living standards is to change the ratio of human effort and capital going into a firm to the amount of value coming out.

By contrast, the objective of a lean transformation is to analyse the core value creating processes of organisations in light of customer needs (which will change), then analyse how to create more value with the same resources so the organisations can grow and society can prosper. This usually involves recasting the firm's product development strategy as well as analysing its day-to-day operating methods.

It's the difference between shifting wealth from one party to another and creating more value, that ideally can be shared with customers, employees, suppliers and owners.

3.12 Information management

You need to manage the way you handle information to make it well ordered and easily retrievable. Developments in electronic data management are becoming easier to handle and time savings are becoming more visible. In many cases, documented records are a part of the customer's requirements. If you are a supplier, your customer may regard a degree of integration with their own EDM system as a strong point in your gaining their long-term confidence.

3.13 Your objective is financial

Most of the factors in this chapter are considered in detail later, with the reasons why you do it and comment on how to. But this chapter has summarised why product development can be such a powerful tool to generate profit.

The only reason you develop a product is to make a profit. The ultimate objective is financial. That's why your product development programme is so important: it sets your company's cost base and your level of future sales. It has a financial impact far greater than you might credit.

Chapter 4
The product development process
A summary of what makes for the fastest, lowest cost product development

Product development is a continual risk-reduction process (See 'Progressively reducing the risk' chart and text after the Preface). You start with the idea and progressively develop its detail. You use a basic template of process steps and shape it, and the way you run and manage it, to give the economic result you want.

4.1 Introduction

There is a generic way of developing products that achieves the best result in the least time, more competently and completely, and at the lowest cost. But no two projects have identical requirements. As Brian Cooke, past Engineering Director of Black and Decker commented:

> *The only way to speed the development process is to construct a unique programme for every project. If you force them all down a prescribed route it is impossible to optimise each programme. Each must have its own programme; there is no common rule that applies to everyone. They are all different.*

You select the best course of action depending on your analysis. So you first generate an economic model to understand what will affect project profitability. For example, as commented in *Chapter 2*, using tight control of your project expenditure as your key project criterion for success can backfire. It can decrease whole life profits by a factor of 10 through delay and missed performance. Such models are dealt with in section 10.4.

4.2 Product development process interactions

4.2.1 The pitfalls

Developing products is a multi-departmental activity. It involves far more than just the technical or marketing area. But few seem to take this into account and, as a

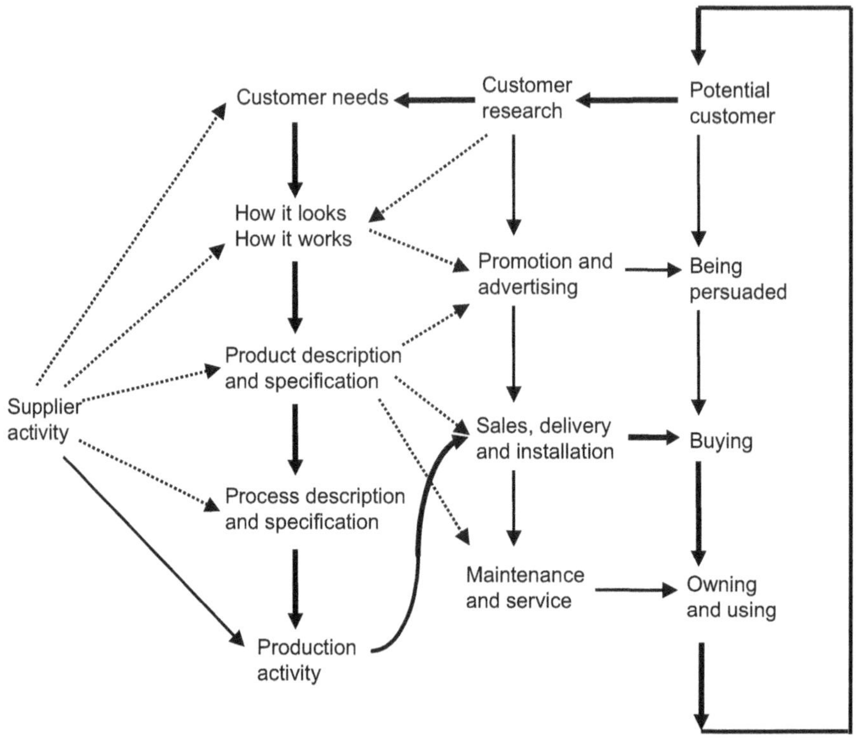

result, few achieve their potential profitability. Their overall business process is 'sub-optimal' to put it kindly:

- First, most companies believe that their documented process governs how they do it when in reality it's their culture; (what you do when no one's looking); culture is set and driven at the top.
- Second, few realise that product development has to be driven from the top across all areas that the product touches, because only the top has the necessary responsibility and accountability level.
- Often, no one operates the company's documented process so staff use the tools but don't realise the potential gain. Often, managers shortcut the process and incur delay and budget over-run through late discoveries caused by skipping early detail. And commonly, staff don't know how to diagnose and solve problems.
- Most have an incomplete or disjointed process, often delegated to a single department, or to too junior a level, or passed from one department to another; as a result, projects incur a succession of late changes that cause unforeseen cost and delay.
- Developing products is a knowledge-acquisition process. Few give enough (or any) time to developing new ideas and recording the new knowledge in a concise and retrievable form, able to be applied quickly to new product concepts. Few consider enough alternative concepts. The result of both is products that lack 'newness' and cost too much to make.

- And last, they put far too many projects in the development programme so the whole lot slows down and they all cost too much. This is because they don't understand the strategic importance of ruthless prioritising at board level to make sure you don't overload your resourcing. Effective capacity planning is rare indeed!

There is a methodology that avoids these pitfalls.

Developing products is a part of a cycle of interaction between your company and the market. It's about company product strategy. Research, design and development follow a logical sequence. You determine the best fit between what customers want and what your company can do, to fit your product strategy. Customer research will (or should) discover what customers want; your strategy makes sure it matches what it's viable for you to do. It's not about slavishly giving customers absolutely everything you think or they say they want: that can lead to ruin, taking more resource and capability than is viable. The resultant products may not be a good fit between your customers' needs and your capability.

4.3 14 principles that minimise time and cost

All the points below are expanded in detail in the chapters that follow. This is the essential summary:

1. **Reorganise yourself to identify waste and reduce it**
   As set out in *Chapter 5*, how you organise your company can double the effectiveness of your spend. And it can generate more profit through a product that customers like better, that is launched faster.
   Don't pass tasks from one department to another; it fudges responsibility, takes more time and causes more rework and drawing revisions. So, for the most complex products, use a team drawn from all areas. For the simplest (a minor manufacturing modification) don't burden the system – get a single department to do it.
2. **Cost your products logically**
   Support your programmes with a costing system that produces the right messages and incentives to your managers. To do this, it must apportion costs in a way that helps to identify wasted effort.
3. **Hurry slowly (*as Socrates advocated*)**
   Don't confuse action with activity. The fastest developers spend far more time and resource on initial upstream stages with company-wide input. This enables them to complete the downstream stages faster to cut total time and cost. By contrast, losers start making bits of product almost before the word *new product* hits their colleagues' ears: this creates a cost-swallowing black hole through leaping before they've even thought.
4. **Base your product's features on customers' needs, not on company opinions, no matter how stratospheric**
   Surprisingly, few companies research their customers. There are effective methodologies you can use. Basing products on *Top Company People's* opinions is a major source of failure.

5. **Spend time considering alternative concepts before you decide on the product's configuration**
   For two good reasons: you'll regret it later if you adopt your initial thoughts. First, you won't get the best, lowest cost, concept first time around. But more important, this is the only phase where you make the time to generate more 'newness' in your product. Companies that don't do this usually produce duller products that lose sales to their competitors.
   There is a fast, formal methodology you can then use that will develop a far better final concept than any you devised initially.
6. **If you don't know that a viable solution exists to a problem, don't try to solve it as a part of your main time-line. And record new knowledge in accessible form for future use**
   If you incorporate new ideas that do not yet have a known production solution, you'll be late and over budget – it's a major reason for over-runs. To avoid it, run a continuous low-cost background development programme where you debug all new concepts and ideas off-line.
   And get as much test information as possible for your outlay. Aim to get failures; they give information that non-failures won't.
   Toyota experiments with simple mock-ups and models of the building blocks of new products. They record on an A3 (see section 9.4.6) the knowledge gained from these experiments and turn it into experience curves. The A3s are catalogued for easy reference by everyone and, through this, new ideas are rapidly de-bugged and applied to new product modules.
7. **When you consider new concepts, try out new building blocks before you commit to taking them further**
   First, don't wait until you build your complete product before you test. If you do, problems will multiply exponentially and the challenge to diagnose and solve them will cripple your time and budget. Test new ideas off-line and debug them before you adopt them as part of a product.
8. **Simulate the performance of the product before you test**
   Don't have a *test–fail–redesign–rebuild–retest* cycle. Simulate the product before you build and test hardware: optimise it on the computer first. The earlier you optimise, exponentially the less it costs you. So do it as early as possible in the concept phases, not after you've detailed it for production release. You will thereby achieve, at least cost, the performance, life and reliability of all your technologies and proposed building blocks. The software to do this is cost-effective; it halves in price every 18 months.
9. **Optimise your concept (see Phase 4) and make and debug simulated off-tool prototypes before you detail it**
   Often the number of interfaces between the building blocks that you've developed above causes the greatest reliability problems and costs. So as early as possible, having optimised the concept, test prototypes with simulated off-tool properties to find and fix these gremlins. Then (where acceptable) make some prototypes for early approval tests; and give some to friendly customers for beta testing.

10. **Don't repeat past mistakes: learn from your history**
Companies persistently repeat mistakes and waste time and money reinventing solutions. There is a way to avoid it (see section 9.3.5).
11. **Understand how to solve problems**
Companies rarely discover the root cause of costly recurring problems. So give your staff formal on-the-job training in problem solving techniques, when and how to use them. Used routinely, they generate large savings.
12. **Your first units' off-production tools are early production units, not prototypes**
Use your first products' off-production tools to confirm what you predicted, to fine-tune your production process, for approval tests where *Phase 4* prototypes were not acceptable, for customer testing, and sell some. They are early production, not prototypes – you built simulated off-tool prototypes in point 9 (*Phase 4,* section 9.4).
13. **Every product development programme will be different**
Design each programme uniquely according to the needs of the particular product. But use a common methodology that you develop continuously.
14. **Run as few projects as possible at the same time**
Cut the number of projects in your development programme by prioritising so you can resource each one effectively. Do not overload the company facility, see below. It will cost you less and you'll get better products and faster.

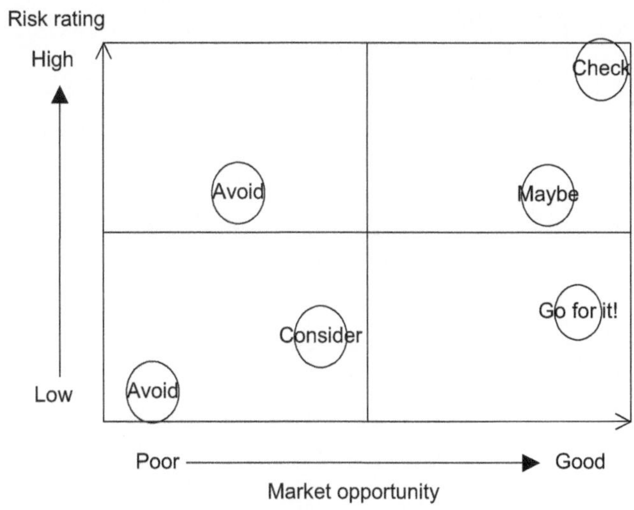

To prioritise, generate a risk–reward plot to see how the benefits of each new project relate to others (see section 10c). Agree each proposal's risk–reward profile by top-level discussion so that everyone agrees the risks and benefits (this is part of the *Phase 1* activities – see section 9.1.3 onwards).

Be sparing (some say brutal) and start only those projects with the best risk–reward profile. If you start only the best, through better control, you will get to market sooner with better products. You will have far fewer change orders and manage them better; with even fewer changes after production release. Plan the resource needed across all departments that have a role including marketing and manufacturing. Allow some spare capacity or you will always be late (the 'spare' always vanishes).

On the other hand, if you take on every project and put too many into your programme, you will have worse solutions through poor staff utilisation from too many interruptions, excessive mental acrobatics, and confused thinking. Postpone starting some so you produce better, innovative solutions brought to market faster at less cost.

Chapter 5
Organising your company to increase profit

Before you embark on developing products, the way your company is organised will either help or hinder. Study this aspect before you start, to understand how much more competitive you could be. Unless you do, you may never develop really successful products fast and at low cost.

5.1 How you do it affects the cost of what you give your customer

You are in business only to make a profit; so continuous cost reduction is routine. You can't price your product at more than the market is prepared to pay, and that price will be continually pressured downwards. So the ways to improve your profits are to reduce your costs and also to move up-market.

Many companies do not understand this. And equally, they don't understand that it is the product development process that generates the major part of their costs as well as their market position.

The combination of poor organisation, route map, understanding and use of tools, and poor programme management causes most companies to spend up to half their development budget to little effect. They increase the product's cost from tasks that increase cost without increasing its value to the customer. This creates unproductive employment.

The key to developing products fast and at the lowest cost is doing the right things in the right way at the right time. Adapting the words from the old song, 'It's not (just) *what* you do; it's (also) the *way* that you do it.' Most of us unwittingly impose unnecessary delays on the programme. Few companies are organised in the best way.

All industry sectors experience related generic problems. So how can we use our budgets to best effect, so we can do all that we want to – and more – within the constraints we have to work within. And how can we do it in the shortest possible time? And finally, how can we be sure that what we're going to develop will result in a product launch that will have more success than we expected? This section will examine how, before we start, we organise ourselves to do it.

Benchmarking sources

- responsiveness
- throughput time
- on-time deliveries
- time to market

} Milliken, Toyota, Motorola

- display
- merchandising
- promotion

} Patagonia

Throughput time

- benchmarked on Toyota and Millikin:

	1990	1994	1995
Rucsacs	18 days	6–8 days	2 days
Garments	6 days	3–4 days	1.5 days

5.2 Benchmarking

Many managers send their staff to another company to see if they can use the lessons they observe to solve their own problems. For example

When starting out, Mike Parsons then the managing director of Karrimor who make outdoor clothing and equipment, visited a number of different companies around the world through facilitated visits. He quickly realised that there was a far better way to run the business. A particular spur was visits to the Milliken company's plants in the UK and the USA. Here was a company that offered to produce carpets to customers' own designs in what appeared to be an impossibly short lead-time. A complete new product, far more complex than anything Karrimor made, would be developed and produced in days. Mike could not believe it was possible: he quickly learned. To find what was possible in industrial measures such as quality and lead-times, he examined operations in companies outside his industry. For display and catalogue, he benchmarked against his own industry. In the years that followed, he progressively moved Karrimor to cut its product development times and costs.

Benchmarking is fine for problems that you can see easily – the obvious ones. You can learn much from seeing how others do it. But you may be able to see and understand only their superficial achievements. It will be harder to diagnose how they got there. Knowing how is not enough; you also need to understand why. And if all you do is catch up with others you won't steal a march on them.

If you discover that your performance is superior to your competitors, you may feel you can relax. If you discover that it's inferior, you may well not understand why. You will tend to compare factors that you work to rather than what they work to. This may hide underlying causes. The important differences are concerned with how they continuously strive to eliminate wasteful activities. How they do this is far harder to identify.

Analysing where you can eliminate wasteful tasks is not the same as benchmarking. You will find interesting information. But in the ultimate, you need to stop worrying about how you compare with others. You'll reach the point where you will become better than the others faster if you have the confidence that you can analyse what you do and develop the fastest and lowest cost way for yourself. By all means see what others do. But, to improve much further, think beyond what you see.

5.3 Improving your added value does not always add value

Creating added value for your company is not the same thing as creating adding value for your customer. Company-added value is sales minus costs. In other words, money you receive from customers minus what it costs you to make the products and run the company. Whereas customer-added value is putting features into the product that make it more attractive and useful to the customer. The two are far from being the same.

> You can create added value for your company but add nothing useful to the product. Clearly this is poor strategy.

Within your company, you have to sort those who are adding customer value from those who are just adding cost. While this is not a manual on general management, you'll find it instructive to examine some general aspects of how your company operates. There is a multitude of routine activities that everyone in your company will take for granted as being necessary. For example you will incur cost through internal tasks that don't contribute to the product's usefulness, such as holding meetings that cost into your overhead. They're always held and no one questions it because they are not identified for what they are: wasted effort. You have to examine activities such as fixing errors, re-design, waiting on others, valueless meetings, reports that are not used – discussed throughout this chapter.

> You have to sort those who are adding customer value from those who are just adding cost.

The challenge you have is to identify those activities that don't add value to the product. They are more difficult to identify than to eliminate. When you identify them, you realise how little of the costly time you fund is spent adding value for the customer.

A simple example is a conversation I had some years ago with a Government official who was supervising awarding grants. He proudly announced that applications were now being processed in only 15 working days compared to 28 days previously. 'How many man-hours does it take to administer one application?' I asked. After some mental churning he replied 'About 5 hours work between 3 departments'. 'So why does it take 15 working days? What happens to it during the other 14½ working days?'

It sounds obvious, but the question is rarely asked. We don't query what works. This is exactly what happens when we develop a product.

5.4 Most companies recycle tasks

Most companies develop products by each department doing its bit then passing it to the next (over-the-wall syndrome). They don't evaluate the whole process first; they become obsessed by planning individual operations in isolation from the rest of the process. This causes tasks to be recycled. Consider a typical sequence of product development activities.

The marketing area or product planners pass a specification (usually incomplete) to the design area that designs and details it. The design area then designs it, taking many decisions with incomplete information. It then passes it to the manufacturing or process-engineering area that discovers there are small changes that could make it easier to manufacture. So they pass it back to the design area to make the changes, and maybe product planning discovers it's back there and updates their specification. It is then passed back to manufacturing engineering ... and this cycle repeats several times. When the production area ultimately receives it, they then start to request their own changes and cycle it back to manufacturing engineering, back to design ... and so on.

> If a department performs a product development task in isolation it's likely to be incomplete.

One company I worked at had a surprisingly elegant method of avoiding this re-cycling. Each area created its own set of marked-up drawings, unknown to the other. So product planning never saw the design drawings; design didn't realise that production engineering had modified their design. And production engineering didn't realise that production had changed their changes. Since everyone was up to their necks in work no one discovered, until some years later when a customer ordered a particular spare part. The service department sub-contracted making the spare part by in order not to disrupt production. They used the design department's drawing. Not surprisingly, the resulting parts wouldn't fit. I leave the ensuing excuse-stream to your imagination.

5.5 Most companies batch the tasks, which queue

Batched tasks waste money. If a department performs a product development task in isolation, the quality of the task is likely to be inferior or incomplete. It lacks vital data from other areas. Transferring tasks from one department to another also loses knowledge. And poor decisions can be passed without accountability because it fudges ownership of who owns responsibility for which part of the product.

If you pass product development tasks sequentially from one department to the next, you waste time. You have to wait until they can do it. This may well utilise all your staff and facilities continuously but it will delay your programme. It will absorb budget and lead-time. Consider it from the point of view of the product being developed.

> Processes need to flow horizontally across the organisation because practically every process touches more than one area, department, function or business unit. And this is often a problem, because the predominant structure of most businesses is vertical. The areas, departments, functions and business units are the primary organisational building blocks as well as the accounting units and career conduits.

First the task is delayed until the department can do it. Then the staff brief themselves to understand what they need to do: wasted effort. Then they do it and maybe need a decision from another department: further delay as it queues for new attention. The new department briefs itself on the problem (more wasted effort) and takes a decision in isolation. You discover later that they didn't know about a vital factor that was known by someone else in another department. So much later, it is 'discovered' and a change is needed (far larger cost of wasted time and effort). The later you make changes, exponentially the more they cost. A change that would cost only $100 at the concept stage could cost $100,000 at the tooling stage.

So what do most companies spend their product development budgets doing? One analysis found that

- 25% was value adding necessary work (e.g. doing the right things correctly at the right time);
- 10% was necessary but did not add value (e.g. reports and travel);
- 30% was rework (e.g. fixing errors and re-design);
- 25% was not working at all (e.g. vacations and waiting);
- 10% was unneeded work (e.g. meetings and reports no one uses).

This is the batch and queue method of working. It can cause up to half your total effort to be wasted.

5.6 Relationship of manufacturing to product development

Culture tends to be uniform across a company. The culture of how you organise and run your production operation interacts with how you run your product development. Unnecessary tasks in both activities waste time and load cost onto your products. You could almost certainly reduce your cost base. For example, a typical ratio of indirect to direct staff in companies in the West is around 1.2 to 1. The best Japanese operations are nearer 0.3 to 1. And most companies use accounting practices that perpetuate incentives to create waste, considered later.

Take a look at your production area. Is there work in progress between operations? Do you make large batches 'to reduce unit cost' because some processes take a long time to change over? Do you have large production stores that feed the final assembly process? And do you manufacture to a sales forecast (always wrong) rather than making a product only when a customer orders it? Is your annual stock turn up with the best at more than 20? Or nearer the engineering average of 9?

> Most departments analyse operational tasks in isolation from the flow of the whole process; their reasoning thereby focuses on the wrong things.

You could be loading your manufacturing costs, using far more floor space than you really need, and be greatly increasing your production lead-time. And you will almost certainly be adding cost and time to your product development by its working to a related culture.

The source of the problem is that most companies' departments analyse operational tasks in isolation from the flow of the whole process. This results in their justifying facilities to perform isolated tasks at high speed (or at all) without costing the effect on the operations that precede or follow it. This causes everyone to focus on the wrong things.

For example instead of fitting options on the line, a world-renowned specialist carmaker used its service department to fit them (company tradition). So the car would be built to standard specification on the line, test driven and passed off. It would then be driven a short distance to the service department to have the options fitted. There, the standard parts were removed and scrapped, wasting untold labour hours and money. They could have been fitted on the line in the first place. And because the standard removed parts were now 'used', they had to be destroyed to prevent re-use (up to $4,000-worth per car)! This destruction reduced the company's potential profit. (Fortunately maybe only 5% of owners requested options!) Exhaust systems, trim, in-car entertainment, wheels – you name it – all received this treatment. As you might imagine, eliminating it would cause a major political battle! I wonder if they're still doing it?

5.6.1 False investment justification

Once you have made a costly investment, or decided on a process, conventional accounting wisdom says that you must justify recovering its costs by using it as much as possible. The more you use it, the lower the theoretical unit cost. But it may produce parts so fast that it creates output mountains that have to be funded, administered and looked after. Few company managers realise that the annual cost of funding stock is up to half its value. (Its analysis is explained in *The Lean Toolbox*, Bicheno, 2nd edition, p. 158.) So, unforeseen, the machine stands idle 40% of the time because it vastly exceeds the rate of sales. Or like the example above, the task maybe entirely unnecessary in the first place.

> The annual cost of funding stock can be up to 50% of the stock's value.

All this obscures the fact that there is a less wasteful method: of working continuously on the task, bringing people to the task rather than dragging it from one department to the next.

5.7 The lessons from the manufacturing area

The evolution of manufacturing systems holds useful lessons that you apply to developing products. In brief, they are as follows.

In the 1950s, Toyota's main requirement was to make dozens of a product, not millions. This still holds; goods produced in small batches satisfy most of our needs. Taiichi Ohno and his technical collaborators at Toyota concluded that the real challenge was to make small-batch production viable by a number of actions that also reduced lead time.

They did this, in most cases without assembly lines, by two developments. First they learned to change tools quickly from one product to the next so they didn't need a huge 'economic' batch size; and second they miniaturised machines and plant to fit into small product cells. Different processing steps (such as machining, painting and assembly) were placed next to each other so the object being manufactured was kept in continuous motion. It was handed from one operator to the next – no conveyor and no pile of part-finished stock between each operation.

Production lead-time was reduced so products could be made only to order, rather than for stock to an (always) inaccurate sales forecast. Costs decreased, and Toyota is still the most profitable carmaker.

5.8 The principles of flow activity

The lesson from the manufacturing area is that tasks can usually be done more efficiently and accurately when the entire task is worked on from start to finish without interruption. You focus on what the task needs, not on your organisation or the equipment you use.

As a manufacturing example, if the total process time to manufacture a product is 10 hours, then that's the aim for the manufacture lead time from production receiving the order. Why spread it over four weeks, with piles of work in progress, under the illusion of optimising the utilisation of your costly plant? It can triple your manufacture cost. Instead, put all the processes end-to-end so the product is in continuous motion. This applies equally to activities in developing the product.

Master the key techniques that identify waste. Start with organising all your processes to flow without interruption. Ensure they are done correctly, at the right time, by the right people.

5.9 Organising to cut waste

In developing a product, the way you do it is as important as what you do. There is a relationship between the two. Much of what you do may add no value whatever.

Can you make it as easy as possible for everyone only to do work that adds value to the product?

> In developing a product, the way you do it is as important as what you do; the key is how you organise to do the work.

The key is how you organise to do the work, naturally, and without undue coercion. Who should contribute input and when; and how does it depend on the complexity of the product?

The first key is how you run your organisation to develop your products: the 'soft' factors. Product development has an impact on every corner of your company's operation. It's company-wide and sets your cost base.

But is how you organise your company's structure really important? Isn't the quality of the people and how they collaborate the governing factor rather than how they are organised? Without the best people, you won't get the best results. But the way you expect them to work strongly influences how effective they are allowed to be. It makes the difference.

Having considered what follows, you may want to revise your organisation's culture a little. It has to be set up so that it naturally focuses on the development process. Most companies are more focused on the way they are organised – departments, functions and areas. This applies to everything the company does, from taking orders through manufacturing to servicing the product. But here we focus on developing the product.

5.10 Flow in product development

Aim to reduce your product development lead time to that of your value adding task time, allowing for the parallel activities, so that lead time equals process time. And aim to eliminate unnecessary tasks.

First get the order of tasks right and only then decide on the structure of each task. Identify tasks that are redundant or that don't add value for the customer. The process (the order and aim of tasks) is your priority. Changing it may well combine or eliminate some tasks. Each should contribute to creating features that your customers will value.

Put yourself in the position of a new product as it progresses from concept to launch. What does it need? For each product family, identify and map, step by step, all activities that add value. Eliminate every step that does not. Companies seldom analyse the *entire* string of activities. When they do, they find staggeringly large amounts of waste. You need the sound basic process structure outlined in *Chapter 9*. When this is in place, you can then identify where waste remains (see section 11.21, identifying process waste).

Such analysis will almost always reveal three types of process step:

1. Those that contribute to value such as detailing a design item.
2. Those that create no value but are unavoidable with current technologies such as having a progress meeting.
3. Those that create no value but are avoidable such as generating a report that nobody will use.

5.11 Being product-led

Drive your operation by product and project, not by department or function. Some of the most effective companies find that the most productive, fastest way is to use a small team representing all areas of their company. For running multiple projects, there are a number of ways you can organise your teams. See *Chapter 10, part b.*

> Drive your operation by a product-led culture, not a departmental one.

The companies with the fastest programmes and lowest costs operate a product-led culture. Product teams look after everything from development to production. Team working and a sound process structure will achieve most of your possible cost and time savings. Effective use of tools and techniques contributes the rest.

The problem is identifying waste; people will naturally eliminate it when it's identified. It's easier to identify where waste occurs by setting up a team to do the whole task because the waste becomes more obvious.

The key is to bring the people to the task in a small core team – it doesn't need to be large. You don't drag the product development tasks from one department to another. The principle is to get everyone around one table. Create a project team with a project manager. Second, call members from all the departments the product touches on its way to the customer, from investigating customers' needs, through developing the product, to making it and delivering it. Members can be part-time. The team manages the task from start to finish. Give them the tools, the support and the authority. They will find the fastest way to do it – monitored to ensure that they meet your company's objectives.

5.12 Changing to a team-based culture

5.12.1 What is culture?

Culture governs how your processes really work rather than how your manual says they should. Work environments of trust, openness and continual learning are built over time. The process is not random but relies on subtlety. Most managers have not been trained to see it. It has to be grown and nurtured; it can't be manufactured;

it needs patience and commitment. Unless continuously mentored, we all tend to apply our mental models and preconceptions rather than from observing what is really happening.

> Culture is led from the top; at its most fundamental level it's what people do when no one's looking.

Analysts find that it operates at superficial and deeper levels: from visible artefacts such as a tools and methods to unquestioned assumptions – what staff do when no one's looking. It's learned through doing, not through study or exhortation. It takes constant practice and relies on personal relationships and mentors. It is mentored from the top down.

5.12.2 Using teams

The fastest way to develop products is to use cross-function teams. Simpler products and changes need small teams. Don't circulate the tasks at arm's length around departments. Define all the people needed and their remit – the vital input of each one; then bring them together to do the task. Generate a route map. Identify and eliminate the steps that add no value.

> The team manages the task from start to finish. Give them the tools, the support and the authority.

A fundamental transfer of power occurs. The autonomous team manages all the routine tasks, helped not hindered by the departments. Each project's progress is monitored regularly (see *Chapter 10*).

But you have to resolve cultural issues ('soft' factors). Using teams with members drawn from different departments can create management unease. Department heads often have a 'vertical silo' mentality. They perceive that losing management control of their staff undermines their traditional careers and power bases. They see it as a threat to their authority, a perceived insecurity to their future, rather than a way to expand their sphere of influence. They don't understand the benefit of team working in expanding the experience and expertise of their staff.

> Engineers are often led by managers who are not engineers themselves and don't understand how engineers work. In the right settings, most engineers will put in long hours of hard work without any prompting. Having them punch timecards in and out, repeatedly cancelling projects, or failing to provide opportunities for professional growth, can destroy their self-motivation.

This has halted many companies in their efforts to change to a team-based organisation. An early 1990s survey of 700 UK manufacturers by *Engineering* magazine discovered that almost 30% had failed to make the transition because of this attitude. There is a workbook[1] that uses an effective methodology developed by doing it with a number of companies. They all made the change successfully – the culture change lasted.

Functional departments control their own budgets, headcount and decision making. Whereas product development can be done in teams without functional boundaries. For some firms this maybe a transition to a new type of culture. The teams, not the departments, do the day-to-day project activities. The functional departments become competence centres to support specific parts of the process. They develop the core skills that develop the company's competitive advantage. Self-directed teams focus on the task and bring all the facilities to the task; the task is not shovelled from one function to the next.

A good example is Toyota, which is a matrix-based organisation with a culture that enables project teams to run the major development tasks. If Toyota's success was the result only of replicable techniques and methods, then many would now be as successful. Their departments act as mentors to assist the work. And the team is not necessarily co-located. It's the pervading culture that makes it work so effectively.

Your peoples' talents are your commercial advantage: knowledge, information and intellectual capability. It's an investment that most companies treat surprisingly carelessly – future capability.

Well-founded teams understand how the whole organisation will gain from any action they take. They manage the whole task from start to finish, with the technical assistance of the departments. With the goals, metrics, resources and training, they determine how best to get the result you want.

At Eurotherm Controls Limited, Dr George Turnbull, Technical Director, Controls Group oversaw reorganising their areas to product teams. He comments: 'To achieve commitment and motivation, it was necessary to abolish isolated departments doing discrete elements of the project and to create multi-discipline project teams, so that all involved had a knowledge of all the issues. The teams, which involve Marketing, Sales, Purchasing, Finance, QA, Test and Production personnel as well as R&D, are co-located in a project area within the old R&D area. The composition of the team varies throughout the project, with non-R&D members returning to their functional roles as appropriate. This dual role for non-R&D members does create problems, with individuals having two bosses. But this loyalty issue is eased by agreement that in all cases 'the project comes first'.

The composition and management of the teams are not cast in stone. They are always up for review and change to find ways to make further improvements to compress the time taken to bring the product to market. We found that in general our previous working procedures were too stringent and top heavy, leading to slowing the development process. We had to have a system that people believed in;

[1] *Using Concurrent Engineering for Better Product Development*, Cranfield University.

it had to be good for the project, the individual and the company. So we simplified and matched the procedures to the need to get the project through on time.'

We value specialised knowledge. But split engineering into too many sub-categories, like product engineer, process engineer, equipment engineer, manufacturing engineer, industrial engineer, etc. and you create bureaucratic turf wars. In a small factory, one engineer may do all the technical work. As company size grows, the trend is to sub-divide the engineering function into sub-sectors; it should be done sparingly.

5.13 The remit of functions (departments) and teams

In a product-led development and production culture, product teams do most of the day-to-day tasks. The remit of the functional departments evolves to support the teams (see *Chapter 12*).

Staff seconded to projects maybe full- or part-time and may or may not be co-located in a core team.

Those not involved directly in projects, perhaps between projects, work back in their parent departments doing development work and training to develop techniques, technology and new ways of working. This can include production workers who become operations specialists able to find and eliminate waste. The departments become the custodians of new technology and knowledge development. They teach team members the new knowledge they develop and research. Typically,

- Technical department staff run the background product development programme: develop new technologies so products can do new things for the user; de-bug new ideas and technologies off-line; develop new materials and methods to eliminate manufacture steps and reduce costs.
- Production engineers develop smaller-scale production devices to fit product cells to enable teams to make products in a continuous flow and change rapidly between variants.
- Purchasing identifies the set of long-term suppliers you will work with; it develops a plan for each supplier to ensure they have the design technologies and production capabilities for the performance you need.
- The traditional quality function combines with a productivity function as an improvement team to eliminate waste by *kaizen*[2] and *kaikaku*[3] activities. The quality area develops a standard set of methods that the product teams apply so every product is right without rework.

[2] *Kaizen* – continuous improvement.

[3] *Kaikaku* – 'instant revolution' – fast, radical change (e.g. AME's *kaizen blitz*).

The fundamental reason for Toyota's success is their business model in which senior managers focus on turning every process into a brilliant one rather than making the numbers and keeping the assets busy in their area.

And in which every manager and employee takes responsibility for solving problems to further improve these processes. A problem-solving process focus drives the efforts of the whole company.

5.14 How teams work

You use a small core team drawn from all the disciplines needed, and get them to work continuously on developing the product.

Teams save effort and time because the multi-discipline multi-function team creates an automatic information exchange between functions. It avoids the need to re-do tasks through discovering too late what the requirements really are. It can reduce late engineering changes by 80%.

The team acts as a permanent core for the project. They collect and process information from all areas, from product planning through all functions inside the company the product touches, to despatch, installation and use.

Decision making in the team becomes team-based and democratic. You set the team up with its own authority and accountability; that is also what creates the team spirit. Without both, the team approach does not work.

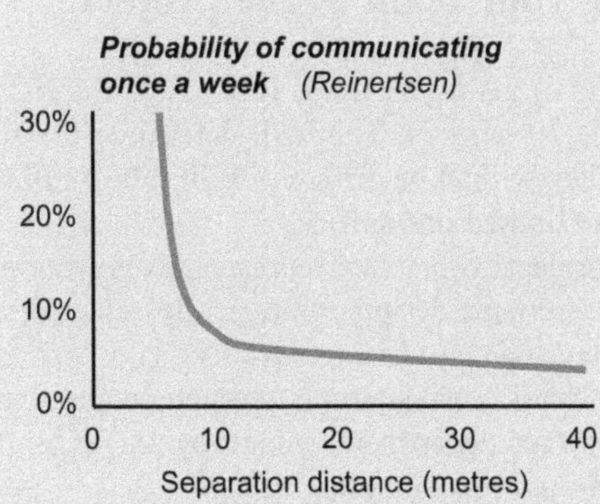

A project manager without authority will not be able to control the project. Being half-hearted or nervous about doing it, such as a partial job with the team not having authority, can cause the traditional delays through department-based priorities taking precedence over the product development task. The old departmental turf wars and loyalties prevent progress. This is a major reason why many companies have not been able to make the step towards autonomous teams and achieve the benefits.

The team should include suppliers that are developing strategically important components or products that you will incorporate into yours. This adds the abilities of your suppliers to those of your in-house team. It helps both you and your supplier: you both develop your capabilities.

If you can co-locate the team, you will eliminate the need for much of the formal communication that was needed before. It makes information exchange automatic and unseen from the people being in close proximity. Staff would previously have had to request information and decisions from other departments. And many problems would not have been recognised through people in other areas not hearing about them.

When Apollo Fire Detectors moved from several separate units to a single building Terry Huppler, their project manager, planned for everyone involved in product development to be within earshot of each other. The company was surprised to discover that, without any other changes, this radically cut product lead-times.

But there is a caveat: through lack of experience, fairly junior team members may approve decisions that their department heads would not. To avoid this, department heads need to take a continuous interest in the progress of each project and tutor their seconded staff. So although the team is autonomous, members must liaise to ensure that they use the knowledge and experience of the mother departments.

5.15 Matching your team to the project

To determine the type of team you need, first assess the risk level of the project (explained in *Chapter 10, part c*). This will determine the size and shape of the development team. They will all be different. Will it be a full multi-discipline team or will a smaller more limited one suffice?

For a high-risk project, you select representatives from every major function right from the outset: perhaps design, strategically selected suppliers, production engineering, manufacturing, purchasing, quality, sales, marketing. But keep the core permanent team small – maybe three to seven people – the smaller the better. Second people from other departments, stage by stage, as requirements dictate. Less risky projects can use less complex teams.

The scope and size of your team will depend on the risk level of your project – all will be different: you tailor each one.

The risk level will generally depend on:

Is it a totally new product to the company (almost starting from a blank sheet), whether in an existing or in a new market to the company? If either, use a full team.

Is it a derivative of an existing product, a modified or new version, or with adaptations or additions? Or an existing product development for a new application in a new market? Here you may want to use a smaller team from fewer departments and perhaps combine some phases of the process.

Is it a simple change to an existing product, or a simple variation to an existing design? You could use a small team centred on the design or production-engineering departments, depending on where the greatest input is needed. You merge even more process phases.

5.16 Setting up the development team for the project

5.16.1 Project champion and review group

The project champion is the single individual who mentors the programme at the highest level.

The product champion is not the main day-to-day driving force behind a project. That is the project manager's duty.

Right at the start of a project, a board member should be nominated to champion and be the voice of the project at board level. The nominee doesn't need to be a product development specialist. In the most successful companies they are usually main board directors with the authority to make final decisions, consulting board colleagues where necessary.

Their responsibility, with the aid of their review team, is to review and approve the resources to support the programme at board level. They remove roadblocks to product development plans, ensure adequate resourcing, and see that bottlenecks are removed along the way. And they act as a mentor to the project manager and the team.

Their first task at a board level project risk review, is to define provisionally the number of formal phases, and therefore reviews, for the programme. This will depend on the project's risk level, considered in *Chapter 10, part c.* They then appoint a review group whose purpose will be to analyse progress at the review points. Chairmanship of the group can change as the project progresses from one phase to the next.

This review meeting is known by several names – project gateway reviews, milestone reviews, phase reviews and so on. The group will review the project manager's report for approval, usually a short tabular document comparing achievement to target. The tasks, and the questions that are likely to be asked at a review, are considered in *Chapter 10, part a*.

5.16.2 Appoint a team leader

As soon as you start to justify the project, you appoint a team leader (project manager).

Following approval to proceed, they are responsible for driving and monitoring day-to-day progress to make sure the project keeps to budget and schedule.

Their task is to see make sure that resource is provided and that roadblocks and bottlenecks are removed. It is an altogether different task from the more strategic task of the project champion. The two tasks should not be confused or combined. If they are, a project champion at team-leader level is unlikely to receive enough support from the top or obtain the resources. Experience shows that the project will not be as successful. Some fail.

The project manager controls the allocation of all resources, human and financial, to the project. If more resources are needed, they are responsible for arranging approval. They keep everyone informed of developments, control the project and are responsible for achieving its deliverables.

The project manager is not a desk-bound administrator. They walk the talk, spending most of their time on the move around the project team. The most effective ones are often technically experienced at the processes they will be controlling, having had a wide experience of operating them. If they don't have that understanding, they can unwittingly make uninformed decisions, slow the project or incur excess costs especially where the product being developed is complex.

As well as technical skills, a team leader needs certain personality traits. Personality profiling can identify the essential characteristics of potential project managers. It is unwise to appoint an autocratic dictator, an oily diplomat or a mouse. There are documented characteristics that are needed and profiling will confirm potential candidates' suitability. See section 5.16.7.

5.16.3 Developing project managers

Developing project managers creates a valuable asset for the business. They can be diffused throughout the organisation as a pool of competence. They acquire a broader outlook; their decisions are better and benefit the business as a whole.

Teams can be set up to have managers whose responsibilities and terms of reference are lightweight or heavyweight, according to their level of authority. A lightweight project manager will have to request authority to do many of the tasks: often a recipe for wasted time. A heavyweight one does not need to; they have the authority and the champion's backing. But you also need the right company culture to make it work.

A lightweight manager may well sit within a department and be responsible to a functional manager. As a consequence the project may experience frustrating delays and not achieve its cost and time targets. The heavyweight manager will be responsible to the board, with no departmental responsibilities other than to run their team – an example of true matrix management. The team may also have performance-related pay based on the product's longer-term performance (never short-term!).

> If a company is embarking on team working for the first time, go straight to a matrix organisation with a heavyweight manager. But you also need the right company culture to make it work.

Typically, a heavyweight team leader (in Toyota, chief engineer) will

- have a broad if not deep engineering knowledge of the product and manufacture process;
- coordinate responsibilities for development across all functions including manufacturing and marketing;
- have a responsibility that lasts for the whole product programme;
- walk the talk among the team, not bury in paperwork or spend much time with formal meetings;
- establish direct contact with customers, prevent the product design deviating from the marketing specification;
- maintain responsibility for specification, target cost, product configuration and major component choice, making sure the concept is accurately translated so that the original customer objectives are faithfully put into the product;
- consult team members on key decisions; listen to and communicate ideas;
- manage conflict and cope with change; strive for consensus but make sure that decisions are taken and acted upon;
- delegate responsibility and define roles; build on team strengths; minimise and compensate for weaknesses;
- mentor junior team members;
- receive reports and report to the board;
- plan and budget, organise and prioritise; monitor and control performance.

Team managers should understand how groups of people work together in reality. A 2002 book[4] on the subject provides insight to team members' behavioural characteristics. It's a useful guide on how these can disrupt progress, and how they can be managed.

5.16.4 Defining team responsibilities

The team must understand its responsibilities. Which are the responsibilities of the functional departments and which are the team's. This boundary can cause untold levels of grief and demotivation if not defined at the outset. Hold an hour or two of briefing, right at the outset, between functional department heads and the team to clarify the authority levels of both.

Generally, you prevent the team from re-inventing routine tasks that are well founded, such as how expenses are claimed or capital expenditure is approved. And you prevent the departments from doing project-specific tasks. They should not be allowed to delay project progress.

Departments are there to help the project achieve its objectives. They assist in training, disseminate the latest knowledge and run the background product development programme, all described in section 5.13. The team runs the project but team members liaise continually with the departments to gain the benefits of common systems, knowledge and expertise.

Unless you give team members the authority to make decisions, you'll end up with endless time-wasting meetings. Hour-to-hour discussion between team members will enable problems to be recognised as they arise, and corrected at least cost. Whereas periodic formal meetings will cause delays through problems not being recognised early enough, incurring far greater cost.

To avoid delay, the team must also be able to function effectively when some members are absent. (See section 5.16.11.)

Do not dictate exactly how the team should conduct each sub-phase of the project. Set their overall targets and let them devise the most economically viable route to achieve time and product cost targets (or whatever your primary targets are for the project). Give advice but avoid imposing straightjackets; they increase time and cost.

And the team must know who is responsible for each interface to avoid split responsibilities not resolving problems. This is because it is interfaces between components and between systems, rather than the number of component parts, that are the major source of development problems.

5.16.5 The core team is appointed

This is done before the concept stage and the team usually remains in place, perhaps in reduced form, until the product is in service. As soon as it is appointed, it can help organise the early stage deliverables, for example the business analysis. This trains the team for its subsequent work and gives it a broader, better-informed

[4] *The Five Dysfunctions of a Team*, Patrick Lencioni, ISBN 0-7879-6075-6.

view. In some cases the team stays together to create Mark 2 and further versions of the product. In Toyota, for example one project team (with membership changes) has existed since the early 1970s to develop the Celica in all its versions.

> You appoint the team before the start of the concept stage.

5.16.6 Small core teams work best

The core team of full-time members does not have to be large. It can range from three or four people upwards to maybe 25 for a large high-risk project. They maybe full time on the project for its entire duration, and have no line-department duties. Other members can be temporarily seconded to work with them as necessary. In some stages of a complex project, temporary secondment may grow the team to a very large size for a short time. For example in developing and installing a North Sea oil platform, teams can exceed 500 in some phases of a billion-dollar installation; but the core team is still quite small.

The big advantage of small teams is their speed of operation. The more people in the team, the more time they will take to arrive at decisions. Discussion time depends on the number of interfaces between people. Indecision time increases exponentially as the team grows in size. So have only a small number of dedicated team members and make maximum use of seconded part-time members.

5.16.7 Putting teams together

Where you have plenty of available staff, you can select on personality traits. The best performance of a team does not come only from technical or management ability. Star teams have the right personality combination, and the members have a mutual liking and respect for each other.

Some general personality requirements need to be incorporated into all team members' job specifications. Members need to be resilient to be able to cope with change, independent and proactive, have good time management and interpersonal skills (equally good for screening new employees). Ideally, you need a combination of personality traits for the team to achieve star performance. This subject has been analysed in detail.[5]

And above all, you need an effective, competent team leader to create a cohesive effort and make sure the team achieves its targets.

Try to avoid creating teams with an 'Apollo' syndrome. When everyone is a star, excessive academic analysis and *not invented here* or *prima donna* attitudes will create conflict that paralyses progress. At the other extreme, don't build a team with inadequate ability that produces shallow, premature agreement on matters that need depth consideration. Nor do you want chaotic personalities that rampage

[5] *Management Teams – Why They Succeed or Fail,* Meredith Belbin.

through problems without systematic methods. You are probably limited in your choice; it's useful to at least be aware of what could happen. Anticipation is all.

5.16.8 Purchasing and its control

This is done better from within the project team, with team members contacting suppliers directly, managed by the purchasing department secondee. Your purchasing department will need to control which suppliers are approved and acceptable to be contacted. They need to rate suppliers on a number of capabilities in order to ensure reliability.

		Potential value of material and components supplied	
Supply strategy		**Low**	**High**
Complexity of what is supplied	High	Close relationship	Strategic partner
	Low	Simple contract	Close relationship

Most companies will rate suppliers on their ability to supply on time and to an acceptable quality and price. But you need more. Suppliers also need to be rated according to their ability to develop what they are supplying you, their product development capability and management performance. Larger companies will have schemes to develop their suppliers' capability and performance.

5.16.9 Suppliers as team members

Those that are of strategic importance to the success of your product should be integrated into the project team from the concept stage onwards. This is key where the supplier contributes a complete system module that they are responsible for designing and proving. It will have to comply with your performance and packaging requirements within the product.

> Suppliers that are of strategic importance to the success of your product should be integrated into the project team from the concept stage onwards.

Their engineer should attend team meetings. In complex projects requiring them to make a large input, ask them to second a member to be resident with your team. This is commonplace in aerospace, electronics and vehicle projects. Treat them like partners not adversaries. Help educate them where they are weak; learn from them; develop a long-term relationship.

Suppliers will need to have their product development programmes running slightly ahead of yours so that product data and early production units are available in good time for you to run and evaluate in your product development programme. Samples must be ready to the right specification and when you need them. Their performance at doing this and their product's performance can form a basis for rating their capability.

5.16.10 Customers

Some key customers contribute to the development process by participating from time to time throughout the project. Don't just show them a *fait accompli* too late to incorporate their suggestions. Hall tests and clinics are often too late – the concept has been developed by that time. In some cases these are just a neck-saving exercise. Consider using some customers as partners throughout the whole project. This will be referred to later when discussing how you research what customers want.

5.16.11 Training for the team

You will almost certainly need to organise training for the project manager or team leader before the team is established. Team members may also require training to learn about the input and needs of other functions involved in the project, as well as how it will be monitored and managed.

> It is vital to train team members formally in problem-solving techniques.

They will also need to be trained in problem solving. They should understand the *concern–cause–countermeasure* routine and the significance of applying the *plan–do–check–act* sequence to all their activities. Managers especially should not be allowed to be ignorant of this and confuse action with activity (proposing solutions before they understand the cause). Routine problem-solving techniques such as cause and effect analysis, 'five whys' and similar techniques should be taught and applied. This is taken further in *Chapter 11*.

Equally important is to train team members to stand in for each other during absences. Being able to use each others' equipment for example can provide useful relief where absence would otherwise delay the programme. This is especially important where multiple projects are in progress and rely on a single facility that can cause a bottleneck.[6] The purpose is to enable stand-ins to do the work adequately, not to create duplicate specialists.

Personnel who have taken part in such teams are usually more useful to the company because they have a wider knowledge of benefit to both the individual and the company. Project managers gain useful experience to prepare them for wider responsibilities. But do not use them as training posts for the inexperienced: that can cause chaos – untold delay and cost.

If team working is recognised and used throughout product development, and training is given, it raises the profile of product development. It tells everyone that

[6] *Managing the Design Factory*, Reinertsen – section on queues.

the company wants to make a wide-ranging effort to continuously improve its products, its fortunes and security, for the future.

5.17 An example: what Malvern Instruments did

Malvern Instruments are world leaders in instruments that analyse particle size. Up to 1994 the company's products were technology-driven. Ideas for products were born and developed within R&D. There was little input from customers or from manufacturing. R&D staff were arrogant enough to believe they knew what was required better than the customer. This could not continue. And new products took 24–30 months to bring to market. Changes in the company's management gave the opportunity to re-structure.

Until 1994, the company's products were technology-driven. Ideas for products were born and developed within R&D. There was arrogance that R&D knew what was required better than the customer.

The new senior management team's task was to align it better to what current and future customers would need. This required them to

- design and develop better products in a far shorter time-scale;
- improve the quality and performance of the products;
- significantly reduce manufacturing costs; and
- improve the support services.

The first objective was to revise their product ranges to use a common architecture. There was little use of building blocks or components. And there had been little interaction between marketing and product development. Both were remote from downstream operation areas such as production engineering, manufacturing and service.

Malvern invested more in product development on the back of growing sales and the acquisition of new technology. This restructuring gave the opportunity to introduce the cultural changes needed. They looked for new skills and attitudes in new recruits – potential future product managers. They needed staff who would understand what the customer wanted, to steer these issues in the multi-disciplined teams. They developed a new management structure:

- Development managers responsible for groups of projects.
- Project managers with total project responsibility.
- Principal engineers co-ordinating disciplines, tools and training.

This was used to identify potential project managers for the future.

By delegating more control, they identified and developed their peoples' capability to manage. Previously a technical expert managed teams who then didn't have time to make their own valuable technical contribution. They were often poor project managers too. Malvern decided that project management should be a role rather than a job. An electronics engineer or physicist with project management skills could now contribute technically as well. As project manager, they had enough overview, technical and product expertise to control the project.

To enable teams to work effectively, the personnel were trained in a range of tools and techniques, including design for assembly (DFA), failure mode and effects analysis (FMEA) and quality function deployment (QFD) (see *Chapter 11*). They were quickly put in place and are now an accepted part of the development process. Team members immediately saw their practical use.

Malvern made the transition to a totally project-based organisation and made it clear that everyone should forget about the previous role of functional departments. Their new role was to provide background functions such as pay and rations, and to organise improvements in knowledge and training.

They are far more responsive to market changes. From 24 to 30 months, they now achieve new products in 12 months. Incremental developments take 4–5 months. Everyone acknowledged the restructuring to be a major improvement. Projects had previously been developed within a product development department. Now it was evident that product development was a company-wide activity.

Chapter 6
Product costing and company costs
Product development sets your company's costs

Most companies cost their products inaccurately. Some companies have wrongly closed operations from doing this. How do you make sure you do it accurately? What are the dangers of under-costing automation?

6.1 What are the costs?

Product costing is a vital part of the product development process. You estimate costs as you proceed, to arrive at the best economic outcome. And you start to commit manufacture cost from the moment the product starts to be configured and its detail defined. It affects such items as

- funding costs of plant and building (the area and volume you need)
  - investment
  - heat, light and power
  - maintenance
- the direct costs of materials and labour
- manufacturing investment in
  - plant and equipment
  - tooling and systems
- manufacturing overhead, such as
  - stock and work in progress
  - process energy cost
  - process consumables
  - floor space costs
  - internal transport
- support costs (depends on product complexity)
  - process design and control
  - systems and computing costs
  - production scheduling
  - inspection and rework
  - purchasing
  - after sales service
- quality and reliability (service costs)

The decisions taken during the product development process set the pattern of your company's cost base for the life of the product.

6.2 How do you calculate product cost?

A survey of 400 or so product developers of all sizes in UK in year 2000 discovered that fewer than 5% estimated the manufacture cost of the product during its development. But 80% calculated the costs after the event – too late to do anything about it (the *You've boobed again fellahs* syndrome). If you find this unbelievable, try the following not untypical example:

A European up-market car maker (then 1000–2000 a year) developed its cars by building a prototype then bringing in contractors (known as CAD-monkeys, such was their low reputation) to help create the engineering drawings needed to purchase parts and materials. At no time during development was much if any costing done, although its line-production sequence was planned in outline. Chief engineers would not brook cost 'interference'. 'My job is to make sure it works and is reliable, not be an accountant.' was the attitude. Despite the cars retailing for well over $150,000, they didn't make a profit. So a cost-down team was created to reduce manufacture cost – well after the event. But it wasn't allowed anywhere near the development teams for new cars! As you might imagine, this produced continuous internal conflict and everyone who had helped develop the original car obstructed their work. And the scope to reduce cost was severely limited by not being able to make radical enough changes early enough, which could have easily been done at the concept stage had the conceivers understood the significance of what they were doing! Section 9.3 shows how you can avoid this.

To calculate product cost, you have to plan how it will be produced. But, in planning the manufacture process, if you use conventional standard costing, or even activity-based costing, you may unwittingly under-cost.

The problem is that if you manufacture faster, standard costing will reduce the apparent product cost whether the output will ever be sold or not. It fails to account for the true costs not only of storing but also of having to fund work in progress (WIP) and inventory, as analysed below. And it fails to account for product that will never be sold and is ultimately scrapped. This is dealt with separately, well after the production event – usually written off at a loss – and not related to the original production costing.

Stock and WIP often occupies more than 30% of the total production area. But it goes unnoticed because it is broken up into a large number of small areas. Indeed, standard costing treats stock and WIP as an asset, assuming that it costs nothing to fund or store, and will all be sold. And standard costing often fails to account for the cost of dedicated plant standing idle.

6.3 The overheads of automation

If you develop a concept that relies on complex production facilities, you are committing the company to fund costly plant, costly management and control overheads, surplus floor space and costly stock holding.

> Departments that make decisions in isolation, without company-wide information, can cause wasteful manufacturing investment and excessive product manufacture cost.

The natural tendency is to use high-cost investment or automation to produce faster. The faster the rate, the lower the apparent unit cost. But there's no point in developing processes to make products' orders of magnitude faster than you can possibly sell them. The low utilisation makes nonsense of the original capex justification, worsens the machine's amortisation rate and can vastly increase the real unit cost. Standard costing will usually fail to notice this. Equally, there's no point in altering the design of the product to permit such a process.

At the start of my career I worked in a large integrated plant (7,000 employees) of one of the UK's largest car, van, truck and tractor makers. In those days no one knew how Toyota operated – Japanese imports tended to be cameras, not yet even motor cycles. The local production engineering development manager at this plant was a real star. He justified buying a broaching machine to cut the internal splines of some large gearbox components. A large pit was dug in the middle of one of the main workshops to accept it, with huge disruption to daily working. Months later the vast broaching machine arrived, and the heavy gang started to erect it. Unfortunately, the top went through the very high and mainly glass north-light roof (oops!). So it languished outside for a few months and was then sold at near scrap price. Not to be outdone, the next capex justification was to replace a bevy of old semi-automated capstan lathes with a multi-spindle chucking auto. This was installed without problem and then proceeded to produce the entire corporation's output requirement in 30% of the working week, with 50% spent resetting it between vast batches and 20% idle. I found this extremely amusing but my manager didn't. The resulting unit cost jumped.

Don't ignore the cost of floor space, minor damage through carelessness to piled work in progress, obsolescence of unsold part-finished product and transport costs to wheel the piled batches around. With excessively high production rate machinery, you create a mountain of work in progress that occupies costly floor space and is gradually used over (say) the next month whereupon you create another one-month's mountain. And because many areas probably use the machine, you can't use small product cells because all their workflows are detoured via this machine.

Standard costing usually fails to calculate the true cost of stock, work in progress and the space, staff time to maintain it and borrowing to fund it, which usually totals about 25% of the stock's value each year.

Companies often do not devote sufficient time and resource to develop how each possible concept might be manufactured. Lack of attention to this detail can result in decisions that are too expensive to revoke later. It usually happens through departments making decisions in isolation without company-wide information. The result may cost more to manufacture, in investment, process time and in control effort. And often the true cost is hidden and impossible to discover because of the way the company organises and allocates its costs between cost centres.

A once-famous European carmaker closed one of its historic marque plants because it was apparently more expensive to assemble cars there than at its far larger dedicated assembly plants. As well as assembling the cars, the historic plant made other things by processes that had a far higher overhead than assembly. But the overhead costs for the entire plant were spread across everything and made the assembly operation look too costly and the other operations look really productive. So they moved the assembly. They then discovered that the overhead on what was left jumped a mile and made what was left uneconomic. So they moved the rest and closed the plant. In reality, it was less costly to assemble the cars there, especially as marque loyalty and tradition made this one of the corporation's smoothest running plants. But the incorrectly allocated plant overhead killed the operation. The site is now a supermarket.

6.4 Simplicity reduces the cost

To reduce product cost, it is more effective to reduce the number of parts and simplify their manufacture than either to automate or to work at a faster pace. You integrate how you develop the product with how you make it, with what customers want and the expected rate of sales. It's a seamless whole. It all starts at the beginning – with product development. The way you do it is symptomatic of how your company is organised.

You fix this at the concept stage by designing the product to be made in the minimum overall lead-time in a small product cell. The key is to be able to change machinery over in only a few minutes to do another operation, or change over to produce a related product. This may require smaller miniaturised equipment to fit into the cell that may have longer individual cycle times than more sophisticated expensive equipments. But it costs less to do it this way than to route the product around a number of separate inflexible production departments, each using costly, and specialised higher-speed equipment. There, they need to process huge batches because it takes so long to change over to another product operation. This increases the funding needed and increases product cost. And it has interesting effects on lead-time:

Traditionally, mechanical products that were machined used to be routed from one group of machines to another. As a rough guide, you'd reckon on processing the batch through one operation a week. The product would spend the remainder of the week sitting in a pile of stock, being transported to the next operation, then waiting for the machine for the next operation to be changed over. So if a component needed 10 operations, that meant a 10-week lead-time. A well-known machine tool manufacturer, that used this system, decided to change to cell production. They desperately needed to reduce their costs and lead-time. Before, they struggled to achieve a product lead-time of 14 months. After changing the whole plant around, the lead-time came down to 6 weeks – still too long by today's standards but a magic reduction.

It took more than a year to complete the change and doing so released funding worth almost a year's sales. A good release of working capital. But interestingly, their costing system still thought the direct manufacture cost of the product was the same as before. This was because traditionally they didn't cost the capital lying on the floor, or the floor space costs, but allocated them to overhead, not to direct cost. So their overheads decreased hugely with the change – but they couldn't figure out why.

Unfortunately, having reduced their lead-time, they discovered that their order book was unexpectedly short because the products had gradually failed to keep pace with what customers wanted. Within a few years the company went out of business. No product – no business; you've got to attend to all aspects to succeed

6.5 How should you calculate manufacture cost? – *from information by Brian Maskell*

6.5.1 The problem with current accounting systems

> Traditional cost and management accounting methods lead to bad decisions for pricing, margins, make or buy and other key issues.

They were designed for traditional mass production and job-shop manufacturing, with measurements such as labour efficiency, machine utilisation, overhead absorption and variance analyses. Traditional costing systems are also wasteful, time-consuming and expensive. They have to be replaced by methods that actively support low waste operation.

6.5.2 The important accounting aspects

It is the business management system that shows you the financial impact of changes. It uses performance measurements that motivate and enhance actions at all levels. These are visual, focused on the value stream, designed to prompt continuous improvement, support work teams and provide fast effective feedback.

Your performance measures become the primary mechanisms of control as processes are brought under control. You eliminate the need for cost accounting and inventory control, which are traditionally needed because processes are out of control.

It does not cut costs but turns waste into available capacity. The financial impact comes as you make strategic decisions on how to use this capacity, and from the cash flow of reduced inventory.

You cost the value stream to eliminate wasteful transactions associated with traditional cost accounting. In return you get timely understandable information.

You make decisions on the profitability and contribution margin of value stream: pricing, profitability, make or buy, new product introduction, product and customer rationalisation. QFD and target costing drives the business by customer value not cost. You don't need to know the cost of a product because all decisions are made by assessing the profitability of the whole value stream, not of individual products or product families. Understand the flow of material, information and cash through the value stream and the bottlenecks or obstacles to flow.

The best way to reduce cost is to increase sales. Use value stream cost analysis to understand where you expend cost with where you create value for the customer and where capacity is available. Use Kaizens to align value and cost through the entire value stream from sales and marketing through product design to shop floor operations. The team studies the customers' needs and develops action plans to increase customer value and profitability. These action plans include changes in sales and marketing, product design changes, operational process improvements and improved administration.

6.6 How standard costing can lead to poor decisions

You need accurate and valid financial information when making important decisions. But you can't do this by using standard costs. It is harmful, misleading and leads to poor decision-making. Use value stream costing instead. Calculate the costs and profitability of the value stream as a whole, not for the individual products within the value stream. Up-to-date information is thereby available on the total costs and profitability of each value stream. For example:

1. **The sales enquiry and the response**
   Company A manufactures hydraulic equipment. The company received a request for XJ2 valves that they have been manufacturing for many years. It is a small equipment and simple to manufacture. The customer, a large distributor in East Anglia, wanted to order 3,000 a month, for years into the future at a target price of £45 per unit. The standard cost for an XJ2 is

Labour time to make an XJ2	750 seconds
Labour cost per hour	£24.73
Labour cost per unit	£5.15
Overhead rate	290%
Overhead cost	£14.94
Material costs for XJ2	£22.33
Standard cost XJ2	£42.42

Because the customer was firm on the £45 price, the order was declined because the profit margin was less than 6%, outside the company's 15% minimum margin rule with a cost of £42.42.

2. **The inevitable outsourcing proposal**
But the sales people were not to be deterred in their quest for a profitable sale. They bought some time from the customer and worked on finding an alternative source. They found a supplier in the Far East that could provide the valve at a significantly lower price than the standard cost:

Price to the customer	£45.00
Cost from the Far Eastern supplier	£30.00
Overhead for incoming logistics	7.5%
Total inbound cost	£32.25
Monthly revenue for 3,000 units	£135,000.00
Monthly cost for 3,000 units	£96,750
Monthly profit	£38,250
Profit margin	28.33%

The sales people told the customer they were happy to meet their price and set about ordering 30,000 units to be made and shipped to the UK from the Far East.

3. **The real costs and the sensible decision**
But the financial controller was not satisfied with the outsourcing decision. The plant had been on a lean journey and he had become a lean leader. He had seen the value stream go from being a long-lead-time, high WIP, traditional production process to truly lean-focused flow. They were still a long way from being really lean – but they had made some progress. The controller reckoned that their value stream could compete with any company in the world. He also disliked the idea of huge inventories and complexity that would come from sourcing overseas. It was the opposite of lean thinking.

So he examined the costs using the value stream costing information reported to the value stream manager each week. Working with the production management team, he found that the additional volume to support the 3,000 units-a-month order would need two more people and two more machines. He worked out how the value stream costs and profitability would change if they added these additional costs and revenues.

	Current state	Incremental costs	Value stream with new order
Current monthly revenue	£1,042,631	£135,000	£1,177,631
Cost of materials	£424,763	£66,990	£491,753
Cost of production employees	£100,464	£7,728	£108,192
Cost of support employees	£208,652		£208,652
Machine costs	£9,858	£939	£10,797
Other value stream costs	£73,115		£73,115
Profit	£225,779		£285,122
Return on sales	21.65%		24.21%

The plant controller then summarised the three approaches: standard cost, outsourcing or making in-house.

	Standard cost - no order	Outsource - take the order	Make in-house - take the order
Current monthly revenue	£1,042,631	£1,177,631	£1,177,631
Material costs	£424,763	£514,763	£491,753
Production employee costs	£100,464	£100,464	£108,192
Support employee costs	£208,652	£208,652	£208,652
Machine costs	£9,858	£9,858	£10,797
Other value stream costs	£73,115	£79,865	£73,115
Profit	£225,779	£264,029	£285,122
Return on sales	21.65%	22.42%	24.21%

Clearly the best course of action was to make the product in-house rather than outsource it. If they made the product in-house, it would make more money and improve profitability. And the company would save itself the problems and additional costs of sourcing from a supplier halfway across the world.

4. **The company's conclusion**
   After seeing this example, senior management concluded:

   a. Making an important decision using a standard cost is misleading and wrong. That you can outsource an item for less than the current standard cost is meaningless.
   b. The team did not abandon the idea of sourcing from overseas, but it was recognized that these decisions must be made using valid and reliable financial information.
   c. Valid and reliable financial information comes from up-to-date value stream direct costs and profitability. These are real numbers, not artificial accounting concepts.
   d. Few people in the company really understood the standard costing system, and using it led to serious mistakes. Value stream costing is simple, readily understandable, and gives information that can be used reliably for decision making.

6.7 Costing as it could be

6.7.1 Summary – costing the product-centred way

> If you manufacture products far faster than you sell them, it increases your costs.

It is almost always less costly to manufacture by a simpler method that matches the rate at which you sell the product. This appears to be illogical because usually all the consequent costs are not taken into the product cost but put to overhead. We should include all the costs such as

- funding the investment in the plant;
- support services and management the plant or automation will need;
- the floor space (including that for storage of stock and work in progress needed to feed the process) and its building HVAC energy;
- tooling, machine consumables and energy to run it;
- rectification and scrap;
- transporting stock to and from the process;
- funding the work in progress awaiting processing.

If you calculate what holding the stock costs you, you will usually find that it approaches 25% of the value of stock and work in progress, every year. Standard costing systems usually ignore this. And you will usually find that to manufacture more slowly and simply, to avoid stock and to reduce the investment in process plant, releases a surprising amount of floor space and working capital.

Every stage of the manufacture process should have a similar production rate to your expected rate of sales. You design each cell to be able to make variants or even other related products. And you design it so you can vary the production rate by using more or fewer operators. And you design the product to be manufactured in this way.

The key is to organise production by product families, then let each product team do its own purchasing, plan its own production and buy its own tooling. This applies to the component and building block production cells (the intermediate processes) as well as to the final process before despatch. The product cells will then incur more than 90% of all the costs involved in making the product. Only a small part will be outside their control, such as the occupancy costs for the space it uses. Using less will reduce costs because the product is charged only for the space it uses. This will produce an accurate cost for the product.

You still produce variance analyses but not based on standard costs. If you stray from the performance target, the team analyses and problem-solves the root cause of the variance. It becomes more obvious where the problem lies because of the team's span of control.

Chapter 7
Product strategy
The starting point – your product strategy

To develop products effectively, you first need to decide what kind of products and which markets you will sell them into. But what should drive your product strategy? How do you decide what to develop? What do customers value and why? How do you plan to satisfy them? What should you do in-house? How should you organise and integrate your suppliers' efforts?

As a starting point to your product development activity, first set out your product strategy: the main strategic issues and what you intend to do to satisfy them. Continually revise it: it's not a straightjacket.

7.1 The products you make

If your company doesn't plan where it wants to be (have a strategy), it may not get anywhere in particular. Developing products is a strategic issue, not a tactical ploy to react to an event. There's no point in learning how to develop and introduce products fast and at low cost if you don't have the capability to supply and service them. Or if you haven't paid sufficient attention to trends that affect sales.

> Update your strategy regularly to reflect market and company changes.

But before we consider product strategy, is it a sensible strategy to move your operations to low–labour cost countries like China. How will this affect your product development?

> No country has ever built a sustainable competitive advantage based on low wages. Wages and other costs inevitably rise. So Chinese firms wanting to export more sophisticated products will learn that quality comes from developing good products made by capable lean processes. The good news is that lean processes do not incur high costs or need complex IT systems.

If you simply move all your operations to, for example China, how can you achieve a sustainable competitive advantage because your competitors will do the same? Real sustainable advantage lies in combining truly lean practices in product design, operations and logistics, purchasing and customer touch with the labour costs at the right location to serve specific customers.

So first you must know who your most important customers are and build feedback loops to help them use your products to solve their problems. This means feeding back to a responsive product development process that gets the next generation product to market in months rather than years.

Second, respond much faster to customer needs. If it takes less than an hour of value-creating time to make the product it should take less than a day to go through the factory. If it needs an hour's work by engineers and procurement to draw up a quote this should take less than a day. Every step in each organisation through which the value stream flows should likewise take a day or less, and little more to flow between organisations. And it should not take more than a few days to reach end customers within your region of sale across Europe. The responsiveness of the end-to-end value stream ought to be measured in days, not months.

7.1.1 Devising a strategy

A product is like a living organism: its performance grows, matures then declines and dies. If you don't have a strategy to renew and evolve your products continually, you too may decline and die. But just setting out your goals as a wish list is not enough. You need a product strategy to understand where you are heading.

How do you arrive at it? First, define your company's aims. Generate a current state–future state map. To derive your future state, ask:

- What answers would you want if you could ask any questions of a clairvoyant?
- If you knew the answers and everything went as well as possible, how would the business differ from now?

Consider, for example what external factors might affect your sales:

- government policies, at home and in your markets abroad;
- what effects could major events (e.g. climate change, wars and so on) have: for example major cost increases of materials, cut supply lines;
- customer behaviour, and the influence of public attitudes;
- science and technology advances;
- competitor pressure to produce cheaper, faster, better, first.

Then generate your response actions:

- define the action gap between the future state and your current one;
- set out the changes you would need;
- define the decisions you would have to make, and when.

Your strategy should be to match your technical capability to what customers will buy. Set out what you are capable of doing now, and the new capabilities you want to develop. To gain and keep good customers you rely on your core competencies, which are

- your team's skills, knowledge and information that competitors can't easily copy;
- your team's creative flair;
- the way you organise and manage your product development, better than your competitors;
- your ability to spot opportunities and get there first.

This should enable you to arrive at

- the type of products you will sell (generic not detail);
- in what quantity, to which markets and by what sales route;
- and when.

What drives company strategy?

Typical excuses for drivers of company strategy are

- an ambition to increase profits;
- to arrest your declining sales;
- to satisfy an individual customer;
- to respond to what the competition has just done (they seized the initiative);
- to respond to what you think the competition may do (but you're not really sure);
- to exploit a neat new piece of technology; or
- delusions of grandeur.

The real drivers are

- what will a sufficient number of customers buy: are the potential markets large enough;
- how you maintain attractiveness and newness across your range;
- how you develop your markets; and
- how you can improve the effectiveness of your company.

If you avoid these real drivers, you may avoid success.

Chapter 8 considers your product programme that will contain a portfolio of projects to satisfy your product strategy.

7.2 Stick to your knitting

Spend resource only on your unique competencies that distinguish your products from your competitors':

- Don't waste your development effort on parts of your product that are not unique to your core competencies, that other companies can do cheaper and better for you.
- Make development alliances with those companies.

Too many companies waste their efforts trying to design and make what they can buy at less cost. Other companies that have specialist capabilities in particular areas can provide it more competently.

For example a small manufacturer that developed specialist packaging machinery had local sub-contractors manufacture most of the components. Success meant that it needed more area to assemble its machines. Its unique capabilities that won the sales lay in the packaging head of the machine. But a relatively conventional conveyor system, which was not a unique competence, occupied two-thirds of the machine's floor space. The solution was to delegate its design and development to a conveyor systems company. The result was better and cost less, because the supplier had far more experience at conveyor systems than the company could ever acquire. The conveyor system was delivered as a complete unit at the final stage of assembly just before each machine was to be despatched. The supplier's staff helped to integrate it with the packaging head and helped in final testing. This took little time and enabled the company to more than double its throughput from the same floor area with a better, lower cost product.

7.2.1 But continually increase your competencies

You can increase your capabilities by strategic alliances with large companies that can contribute abilities on a scale that you can't even contemplate. You may find one with technology capabilities that can work with you to develop a unique building block to make your product far more useful to the customer.

> Don't waste your valuable resources on planning to develop those parts of your product that are outside your core competencies; let others do it in partnership with you.

TSS Limited is a small company that develops world-leading marine instrumentation and systems, such as for sensing motion, heading and positioning. More than half their products' content is designed and developed by key suppliers. But they retain and work on all their products' core technology themselves. They carefully project manage co-development projects with key suppliers. For example, British Aerospace is a major component supplier fundamental to some of their products. They are part of the TSS product development team. 'The relationship with suppliers has to be mutual. You give them information and involve them in developments up-front. It's a partnership effort and it pays real dividends.'

From evaluating what your core competencies are, and what other companies' competencies you will make use of, you can set out the kind and extent of products that you will develop. You test your product ideas and suggestions against this. If you want to deviate from this strategy, you have to plan how you will acquire any additional competencies needed.

7.2.2 Technology road mapping

In this age of information (the Internet, staff movement from company to company), technological advantage is the least sustainable of any advance. Nonetheless you have to continuously develop your ability. Those that do it best can lead the market. Those that don't bother may milk their cows in the short term, but long term will have to find another way.

> Technological advance is one of the least sustainable but is nonetheless essential for survival.

You can help your product programme keep to schedule while adopting new technologies by using so-called technology road mapping. This is a simple adaptation of the QFD (quality function deployment) process. It doesn't need rocket science to do it (despite what some consultancies may have you believe). In your matrix, instead of customer needs versus design features, you list future products versus the new technologies they will need, with the date by which they will be needed. In that way, you don't get caught out by discovering too late that you can't actually put your product plan into action because of missing technology competencies.

7.2.3 Your design and technology level and your strategy

Continuing with tried and tested products is safe only if customers continue to behave in the same way. They rarely do. So when competitors develop a better way to satisfy their needs, your sales decline. Many markets are fickle; many are fashion dependent; some are a softer touch.

> Playing safe, continuing with tried and tested products, is only safe if customers' continue to behave in the same way; they rarely do.

You can classify products by the newness of their technology and design in a product pyramid, suggested by Gunther Krüse and others ('A Nation of Sweatshop Keepers', *Management Today*, April 1997). They classified them as

- Innovative new technology products, for example molecular vacuum pumps.
- Existing high-technology products, for example specialist machinery, aerospace.
- Commodity products, for example white goods, cars, standard computers.
- Basic components, for example standard springs, fasteners, household wares.

Class 1 and 2 higher technology products are at the leading edge of technology or innovation. Their demand renders them insensitive to price. These companies spend highly on product development and on acquiring the background technologies on product or process or both. They can spend 20% of sales or more to achieve

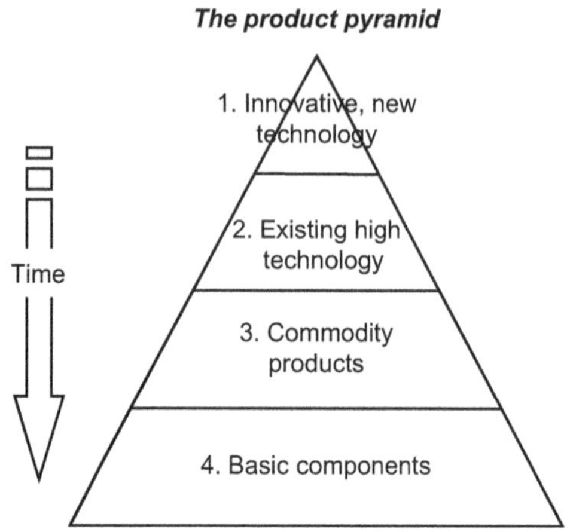

innovative design and new technology content, with high margins and return on investment.

But in time, as the technology becomes better known and less costly to apply, a Class 1 product will be pushed down the pyramid. So if companies are to remain in Class 1 or 2, they continually have to develop products with new design and acquire new technology capabilities.

The lower the technology content, the more difficult it is to compete with newly industrialised countries; and the more important exchange rate and process efficiency becomes.

In lower technology Class 3 and 4 products, lowest cost and best design are vital. For example in the automotive supply base most products are manufactured in large quantities by a number of suppliers. Here, you not only need the best design skills but also see that processes are as free from wasteful activities as possible to minimise cost. The most profitable produce only when a customer orders, and have little or no stock or work in progress anywhere in their facilities. See Womack and Jones, *Lean Thinking*.

The lower the design, engineering and technology content, the more difficult it is to compete with newly industrialised countries, and the more intense the competition. Nearly any country can make basic components such as standard engineering fasteners, plastics household ware and the like. The days are numbered for the many Class 3 and 4 companies in the West that do not run zero-stock[1] operations (Toyota production system).

[1] *Non Stock Production,* Shigeo Shingo.

In the United Kingdom we should be concentrating on categories 1 and 2 up-market products because that is where our inherent abilities lie.

To grow manufacturing margins in the United Kingdom we should concentrate on Class 1 and 2 products because it's where our inherent abilities lie. Such up-market products are less price-sensitive, create higher margins and have the potential to generate a far higher return on investment. We would be affected less by exchange rate instability and high costs.

This should be a major consideration in setting your strategy. If you want to survive in Class 3 or 4, you will need a very different approach from in Class 1 or 2. And you will have to accept that your main thrust to retain margins will be price competition, often against countries with far lower labour costs than yours. A challenging strategy; some say an impossible one.

7.2.4 Should you give customers exactly what they want?

Markets and products inter-relate. Some products have a fairly short life and the customer may be looking for something new and unique. Others may be very stable but more sensitive to the quality of the offering.

There's no point in learning how to develop and introduce products fast and at low cost if customers don't want to buy enough of them.

In the 1960s and 1970s, Japanese motorcycle manufacturers displaced the UK manufacturers from their home market. With the increased ownership of cars, motorcycling was becoming more a leisure pursuit than a transport necessity. New customers, who may not have previously considered a motorcycle, were attracted to machines with more than just the basic necessities. It started to become a life-style purchase. The UK manufacturers did not research potential customers; they still catered for the traditional purchaser who wanted basic transport. Their traditional customers deserted them for cars; new customers bought Japanese imports – and the UK manufacturers didn't understand why either did so. Instead they thought that the new machines were unnecessarily complex, too heavy and too cosmetic. They had not followed what the Japanese were doing at home or why. They did not consider their strategy and had far less investment than the Japanese in the manufacturing plant or in product development. They tightened their belts and reacted too little and too late. By the end of the 1970s, the UK industry had all but disappeared.

There is a difference between developing products simply to please particular customers and aiming to meet a market requirement.

Product strategy is what it means, so don't follow the whim of every customer and get your product programme dictated by your customers' tactical wants. This may look sensible at first. But there is a difference between developing products simply to please particular customers and aiming to meet a market requirement. The first will have you producing variants galore at their every whim. You quickly get an over-complex product range, costly to manage with a high proportion of unprofitable products. And there's the insecurity of tying your fortunes to too few customers. That is not the meaning of *being customer-led.*

7.3 Doing nothing may be high risk

Many manufacturers take a conservative approach, preferring the low-risk route of channelling new or revamped products into existing markets rather than methodically researching and entering new ones. Much of the reason for taking a safe route stems from a lack of confidence on how to investigate customers and new markets. This needs to be addressed if you want to win new territory. Section 9.2 gives guidance on this. If you don't, others may gently displace you from the only market you have.

To succeed, you define your target and aim at it. This applies as much to improving existing products as to developing new ones.

> Doing nothing is a high-risk strategy. It's a conscious decision, even if it's informal, unwritten or even unwitting.

Often companies don't plan their future and product developments are not done to any strategic target.

To use a motoring analogy, not having a strategy is akin to taking the least congested route at each junction, without a thought of where it would lead. After a time, you probably won't know where you are.

Many companies do this – what comes easiest without aiming at a destination. Short-term reactive decisions look easy at the time. They react to an opportunity without thinking through the consequences. For example they may react defensively to what competitors are doing, after they have done it, too late for an effective response. Their objectives amount to little more than trying to keep their staff fully employed. Many often don't even succeed at that.

Companies that don't plan their future direction can lead themselves away from their core capabilities and into blind alleys. Some plan new products that need large deviations from their competencies. There's no point in being able to develop new products fast, at low cost, if the products are not planned to give you a strategic advantage.

7.3.1 Plan your route

You need a product portfolio of mixed risk-level projects. Your immediate aim is to generate cash to support your operation; and then to fund the more innovative longer-term projects that assure your future cash generation.

> If you bias your programme towards short-term results, you may never achieve the really profitable long-term winners.

Totally safe projects rarely make the largest rewards. But that doesn't mean that you ignore all high-risk projects that have a high-potential return. You can manage the risk down as you proceed. A risk that may seem almost unacceptable at the outset may be able to be converted to a risk-free outcome. The key is how you phase and control your individual programmes, and manage the whole process from end-to-end. How you assess and manage risk is very important; it's dealt with in detail later.

You analyse where you want to be, set targets and plan a staged programme to get you there – a detailed route to your destination. You plan new products to fulfil strategic aims, to make progress along a planned path. You generate a product portfolio to achieve your long-term strategy.

Once a product development programme has been decided upon, the way ahead is relatively straightforward. Deciding upon and defining the projects is the difficult part, especially if the company prides itself on continually launching creative or innovative products. Again, dealt with later.

The strategy in practice

7.4 Your ideas-generating mechanisms

What mechanisms do you have in place to make sure you glean every scrap of useful knowledge that is being generated worldwide in your field?

> Having set out your base product strategy that defines the kind of products you will develop, use it to direct your idea-finding process.

7.4.1 Finding viable ideas

A few companies are bursting with ideas; they have problems selecting the few that will be the most successful. Here, the important aspect is to select only the most viable projects. But selection is often irrational: how do they do it? Is it gut feeling (randomly right or wrong), pet ideas or logical analysis of customers' wants combined with some inspiration? Does it match their product strategy? We look at charting risk against reward later.

Other companies are almost bereft of ideas. Here, the problem is to guide them to realise that they can generate ideas systematically. There are numerous routes to doing this.

7.5 Generating product ideas

You can't have too many new ideas; doing what you have always done can lead to ruin. Figure 7.1 shows a typical process map for capturing them. The result is fed into the Phase 1 screening process (section 9.1).

7.5.1 Internal company ideas

There are many ways of encouraging ideas generation; suggestion schemes are a traditional one. But beware of presenting and handling it in a way that demotivates staff from using it. In many companies it's regarded as a joke because of lethargic response or mean awards. In the United Kingdom, only one company in eight has a suggestion scheme, many times less than in the United States. As a result, only nine innovative projects are generated each year for every 100 employees in United Kingdom compared to more than 30 in the United States.

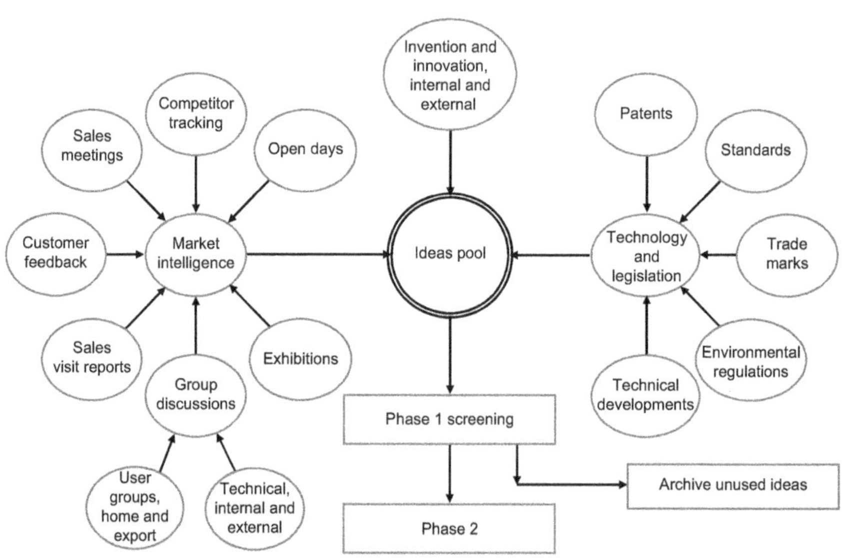

Figure 7.1 Typical ideas-generating process map

> You can't afford not to use a range of resources to generate ideas as feedstock for your initial screening.

The most innovative ideas tend to come from cross-function teams. Most companies don't take advantage of this source of ideas. Quality circles, continuous improvement groups and equivalent schemes are a good source. They need to be well organised and visibly supported by top management who need to be seen to participate. And you make sure your product strategy is known to all, so the ideas generated are relevant.

7.5.2 Ideas from customers and agents

Do you capture continuous feedback from customers who are using your products? They know how they want to use them better than you. And they know what they like and dislike, what they value and don't value.

Customers modify or adapt products to new applications – a host of fertile ideas for variants and new products. Your field staff can monitor this through, for example structured pro-forma visit reports.

> In some industries, 80% of all new product ideas come from customers' modifying and adapting existing products, whether yours or your competitors'.

But few companies use this source of ideas. For example in year 2000, the European car industry had virtually no mechanism to do this. While they conducted costly surveys to monitor dealer performance and what went wrong, virtually none surveyed what design features customers liked or disliked in the car they had bought. Some carmakers are still losing significant market share through their 'I know best' attitude.

But beware of acting only on sales representative feedback, treated as gospel. Their requirement tends to be motivated by short-term tactical considerations (e.g. their sales bonus) rather than strategic. It needs to be filtered.

7.5.3 Ideas from suppliers

Can you make strategic use of key suppliers' developments to incorporate into your own products? See section 7.8 on how you might do this.

7.5.4 Personnel policy

Are you recruiting the best do-ers as well as managers? Who sets your recruiting standards? People bring new ideas and experience from other companies.

7.5.5 Licensing

Could you license ideas to use in some areas of your product building blocks? Buying licences is often less costly and faster than doing the research yourself.

7.5.6 Systematic desk research

You need a routine mechanism to glean every scrap of useful knowledge that is being generated worldwide in your field. Continuous screening of literature and patent applications can provide useful ideas. You can often sub-contract this quite inexpensively to local part-time library researchers to take load off your own staff.

7.5.7 Exhibitions

It's amazing how indiscreet sales staff can be at exhibitions. They will probably tell you their confidential projects not yet announced – in a kind of macho 'we're the best' way. Can be an excellent source of competitor intelligence. Make sure your staff don't do the same!

7.5.8 Think tanks and ideas groups

A group of the company's senior members meet, sometimes with outside specialists and opinion formers, to pool ideas on what is, and is not, a good idea for product development. Discussions usually move in many different directions, and often yield interesting results.

You can assemble groups of users who are renowned for their expertise and have them meet two or three times a year. For example Karrimor would hold periodic meetings to get ideas from international mountaineers and trekkers, as well as from staff who made extensive or exceptional use of their equipment. Ideas emerged from discussing trends and needs.

7.5.9 University research

Ideas come from many sources. Liaisons with particular specialist academic or research centres are often fruitful. The progress of basic research often reveals new materials, new processes and new instrumentation, many of which have direct industrial applications.

7.5.10 New production technology and methods

New production processes and technology will often reveal the answer to adopting a previously unviable product idea. Lasers made it possible to pierce small ($\varnothing < 0.01$ mm) holes accurately at high speed. Applying genetic engineering to new materials will lead to advances in areas so far unknown. You need to have a method of monitoring such developments; an external researcher can often do it for you inexpensively.

7.5.11 Directories

Addresses and telephone numbers may be obtained in the United Kingdom from the DTI and from the Engineering and Physical Science Research Council (EPSRC) and from some trade associations. The European Union has a website where you can access ideas that have been developed to the feasibility stage and are seeking a company to exploit them commercially. Use the search engine on the EU CORDIS website at: http://cordis.europa.eu/projects/home_en.html.

There are many printed directories that lead to lists of ideas that are ready for exploitation.

7.6 Why do customers buy your product?

7.6.1 Comparing yourself with competitors

You should know your market share in the areas that contribute your major profit. Four in ten don't know theirs.

Compare the competence and level of your product development activity with that of your competitors. Do they achieve faster, less costly programmes? Are their products less costly to manufacture? Do they introduce products more or less frequently than you do?

When you analyse your competitors' products, to discover their features and technologies, don't assume that what they do always right; often it isn't. And there may be a logic that applies to their product range but not to yours.

For example in the 1960s a multinational carmaker spent a large tranche of money developing a range of V6 engines because they discovered that a major competitor was planning to introduce a range of V6 engines in their cars. What they hadn't taken into account was that the competitor was also introducing other vehicles that would use a far larger volume of V4 engines. It commonised most of its V6 components with those of the V4s, which made the cost of the V6 quite low. After two years of costly product development, the first company scrapped their programme because they could not make their V6s cost-competitive.

The conclusion is that it's more profitable to devise your own strategy based on sound reasoning, rather than copy others. You may never know why they do what they do.

7.6.2 Perceived value

A customer's perceived value of your product bears little relation to what it costs you to manufacture it. Customers neither know nor care. Despite this, the first temptation is usually to find a cheaper way of making it. The second temptation is to add features and elaborate.

The real solution is to analyse what the customer values: the only reason your operation exists is to create what the customer will value at an attractive price. And the product development process is the primary mechanism to achieve this.

Don't strive just to be the cheapest by planning to develop the most basic stripped out specification to cut cost; you'll find very few will want to buy it.

Customers rarely buy the lowest cost product. They first gather information on products that meet their needs then decide between them. They buy on value at the specification level they want, not the cheapest irrespective of specification. This impacts your strategy.

Use your superior technology and your knowledge of what customers perceive as valuable. Deciding where to pitch this will set your profits. And will you make huge quantities at a low margin or is it more viable to make smaller quantities at a high margin? You can do both (see L&G in section 7.7.4).

Two European knitwear manufacturers re-equipped their factories with very costly state-of-the-art knitting machinery. Manufacturer-1 used their machines to produce huge quantities of basic low-cost knitwear for chain-store retailers. They went bankrupt because they couldn't produce them as cheaply as a competitor from a newly industrialised country that used low-cost labour to operate less costly more labour-intensive machines. By contrast, manufacturer-2 produced complex knitwear for exclusive fashion outlets using costly wools. They have gone from strength to strength and make better profits every year.

7.7 Interaction with manufacturing strategy

7.7.1 How do the two interact?

Competition increases relentlessly. Factors such as the cost of labour create real problems. But often companies don't appreciate the advantages they have. In 1999, for example the employment costs of labour in the United Kingdom were the third lowest of any EU country, despite the pound sterling's recent appreciation. But despite that, most CEOs were lamenting their cost uncompetitiveness compared to their continental neighbours'.

> To increase profits you continuously reduce costs because the customer decides the price of your product.

Sudden currency movements do need committed action to counter. But over-reliance on sub-contracting components to a lower cost region can generate more problems in the longer term than it solves (see remainder of section 7.7). It is better first to examine your own operation to identify and cut waste. See Womack and Jones' *Lean Thinking*.

You reduce your costs continuously to continually improve profits because customers set the price you can charge. This has to be by a combination of product and process design. If your competitor sells a product with equivalent features to yours at a far lower price, you have to discover how.

But is price advantage the sole reason why customers buy your product? If it is, low–labour cost countries will ultimately take your market. If you have large sales in remote markets, would it be better to manufacture those products in the remote markets? But in the longer term, would this mean that you would not be able to design for manufacturing processes if you did not manufacture yourself (see options in section 7.7, below)? It is far better to develop products that are more up-market that you can afford to manufacture.

7.7.2 Should you move manufacture to low labour-cost areas?

Don't be seduced solely by the prospect of lower costs of a module or component – there's more to it. Moving manufacture to other counties is potentially vulnerable to local conditions beyond your control. Production may be too remote from your customers. It lengthens the supply chain and ties up mega-cash in storage of bulk deliveries. The annual cost of storing and managing stock is more that 50% of its value. So every year, stock ties up cash of 1.5 times its value. Consider the management costs you will incur to ensure consistent quality levels and deliveries. How will you avoid large defective shipments that could be time-critical to your production?

How does your product strategy interact with your manufacturing strategy? Consider the possible implications of moving production to remote low–labour cost areas: it doesn't always pay – there are hidden costs.

And if manufacture is far from your markets, transport costs are significant, dependent on product size: for example in 2005 it's cheaper to make computer printers in the European Union rather than the Far East if more than 7% needs freighting; mobile phones – ditto if more than 12%. Consider the effect on your supplies if an energy or climate change crisis doubled or tripled transport costs.

As low–labour cost countries expand their economies and become more affluent, their manufacture costs will rise, as Japan's did since the Second World War. Offshore manufacturing in low-cost countries will increase sharply as companies there develop capacity in industries previously unaffected. They will also develop their own product development capability and may supplant your sales. By the time that occurs, will you still possess sufficient manufacturing expertise? It is strategically important to control your manufacture technology as well as your product development.

It's logical to manufacture in a remote market if the bulk of your sales are there; but staying local is preferable when labour inputs are low, product life cycles short, and obsolescence costs high.

However, not all of the developed world's manufacturing jobs are destined to move to low-wage countries. Some manufacturing is best kept local, even in expensive locations like California. At least two-thirds of that state's manufacturing jobs are in capital-intensive or customer-service-intensive industries that would benefit little from off-shoring to cheaper production locations.

Staying local is preferable when labour inputs are low, product life cycles short and obsolescence costs high.

7.7.3 How can you cut your manufacture cost in the United Kingdom?

If cost is critical but not the sole factor, how much could you cut your UK manufacture costs? Totally automate production: what cost £50,000 ten years ago now costs £5,000. Reduce costs – apply Toyota production system methodology to allow you to make only to order, in a very short lead-time, and eliminate stock or work in progress. This radically cuts labour, floor space and overhead.

Redesign the product to eliminate labour content; for example Swatch redesigned their watches to reduce labour content to less than 5%. The Japanese instigated their continuous cost-down, features-up programmes decades ago in response to the Yen appreciating for more than two decades. It rose gradually at an average of 5% a year from 270 to almost 100 to the US$.

7.7.4 Consider alternative manufacturing strategies

There are alternative strategies to use low–labour cost regions without threatening your competitiveness.

You could consider producing only your high-value variants in the United Kingdom and manufacture the lower-value high-volume products that you sell to remote markets in locally funded and managed licensor plants. Arrange that they depend on your product development capability to update the product regularly; and possibly supply a few small high-technology parts that are beyond their capability.

For example Landis and Gyr (L&G) are Swiss manufacturers of electricity metering equipment. In the late 1960s, they developed a policy of moving the production of their high-volume low-margin products to within the lower–labour cost markets that needed to buy them. They set up locally owned, financed and managed companies to make them under licence. But they retained ownership of the product design and manufacture technology, which they updated regularly. They also supplied certain high-precision items that they could make cheaply in quantity but which required know-how investment that was only justified in a central facility. These items cost very little and made little impact on the local value content. By this policy, of designing all the products and keeping the manufacture only of the most complex and least price sensitive in Switzerland, they avoided the need to sub-contract to low–labour cost areas and retained products and manufacture know-how at home. They also avoided other countries competing by developing their own similar products.

Or could you virtually eliminate labour cost, for example by small automated facilities nearer the point of sale?

7.7.5 Use product development to cut cost

Focus more on what you manufacture rather than on the cost of making it. Revising the specification and making the product simpler to manufacture could cut out more cost than any other area. Generate competitive advantage through developing your product and manufacturing technology together and develop rather than lose manufacturing expertise. Collaborate with suppliers and research providers.

Around 80% of your company's cost base is created during product development. It determines your production philosophy and costs.

You need to know just how much each product contributes to the bottom line. And within that, the cost analysis of each one. A Pareto analysis identifies the highest cost items to tackle first within each product.

Develop improved up-market new products faster than the competition or low–labour cost markets can consume or replicate. Devise products with technology knowledge or protected intellectual property they can't compete with. This is a large opportunity: China's niche quality goods market is as big as Europe's mass market – they have 25% of the world's population.

Make sure your potential customers value every feature of your product and don't give them any they don't value (see *Chapter 11*, QFD).

For example the 600 Group Machine Tool Division developed their Tornado CNC lathe in the early 1990s from analysing in detail what customers used their lathes to do and what they did and did not value. They discovered that many of the features offered by their higher cost competitors were not valued. They also discovered that other useful features, which would cost very little to add, were not offered. They concentrated on simplifying the new machine's features by offering only what the customer valued, and its design for simpler manufacture. The result was a machine that was as accurate and productive as their most expensive competitor's but priced at 30% less. Within five months of the launch, they achieved a 22% market share.

7.8 Operational considerations

7.8.1 Incremental development – your products' age profile

You need to introduce new products frequently to maintain or increase market share, not wait until the decline starts then introduce one as a defensive measure – invariably too late. Does the sales success of your products relate to their newness? If it doesn't, this may identify new possibilities in your product strategy.

The most profitable manufacturers tend to have 80% of their sales from products that are less than 5 years old.

An easier way to introduce products more frequently is to use some incremental development in your programmes (see *Chapter 11*). Small changes, very visible to the customer, are introduced more often, based on the results of continuous customer research. It reduces risk and takes less time and cost to develop the product. It makes it easier to launch a product to meet a particular event such as an exhibition. But first your products' architecture needs a logical structure and common systems and components between products (see *Chapter 8*).

7.8.2 Is centralised development best?

Larger companies that sell products in many continents often cut costs by centralising their product development activities in one country. But this risks losing touch with the needs of local markets. To combat this, some have development centres in, for example four major world areas. They do it because cultures in different part of the world need individual versions or different products.

> Cultures in different parts of the world need different products.

The Machine that Changed the World[2] describes how the car industry practises this. Each centre imports some components that are developed centrally. And the products it develops to sell in quantity in one market can often be exported in small numbers as a higher price specialist product to another. So each product centre exports to its other areas, swapping products for components. This helps smooth currency fluctuations.

7.8.3 What you learn from others

To retain viable sales volume in the long term, you can't continually retrench or you'll end up with insufficient sales. You will want to improve, or at the very least maintain, your market share. As part of planning your strategy, you may investigate all aspects of the performance of your competitors in the markets you are attacking.

> Continuous retrenching leads to corporate anorexia.

But the most interesting lessons often come from companies outside your sector that have world class performance in particular aspects of their operation. Don't expect to find one company excelling in all areas. Particular companies may excel in a particular task area, in a non-competing sector. Visit them to discover what they do; chief executives are often quite flattered to be asked. But be aware that visiting other companies is just for idea gathering and motivation. (See section 5.2.)

[2] *The Machine that Changed the World,* Womack, Jones and Roos.

It's not a substitute for having a policy to analyse everything you do; every task contributes to the value of the final product in the hand of the customer.

7.8.4 Effect of product strategy on sales growth

Analysis[3] has shown that strategic portfolio planning should aim at using common components and technologies across several products. You overlap projects and design the components to be used in more than one product. This is a key strategy of how you operate (see *Chapter 10, part c*).

> Transferring technology between products has by far the largest effect on market share and sales growth.

7.8.5 Share your strategy with that of your key suppliers

As part of the process, it can be highly profitable to share information with suppliers that are strategically essential to your products. Few companies do this, but it can make a big contribution to developing better products and cutting future time and cost. It is a confidential exchange of information on your and their future direction, to see where developments can be mutually useful. You will often find that suppliers have new developments that can catalyse your ideas. Or you may find that your supplier is trying to be helpful but is proceeding in a direction that is not useful. In that case, they may be grateful for some reorientation.

Identify the *close relationship* companies from the table. These have the greatest benefit to your programmes

- complexity and value of their development work (see table);
- their current performance;
- when the supplier would benefit most, based on the timing of their involvement in the next development project.

		Potential value of material and components supplied	
Supply strategy		**Low**	**High**
Complexity of what is supplied	High	Close relationship	Strategic partner
	Low	Simple contract	Close relationship

From each supplier, identify who you would like to participate in your development team.

[3] *Thinking beyond Lean*, Cusumano and Nobeoka.

It can be strategically profitable to share information with suppliers that are essential to your products.

As one leading industry figure commented: 'Long-term relationships will always beat competitors that use abusive relationships for short-term gain. If you want innovative and advanced technology ahead of your competitors, you need motivated suppliers with a relationship developed from mutual trust. Trust does not come from statements on what you tell the outside world your policy is. It stems from what you do – the reality of your actions.'

You both gain advance knowledge of each other's plans to benefit future products. This might be an advance in materials, in particular components systems or in technology. Or about the availability and price trends of future purchases. And you gain advantage by planning to develop in directions that neither might have considered: competitive advantage through continual strategic collaboration.

7.9 Devise and communicate your product strategy down the organisation

Few CEOs claim to be happy with the way their companies devise strategy.

The few that do update their plan say that the annual strategy-planning process yields few good ideas and is often fraught with politics.

There is a better way. A study of more than 50 large companies suggests that the annual planning get-together works best to communicate and prepare managers for strategies made at other, less formal times and places. It is not where you develop your definitive strategy.

Companies perform better by improving ways to develop it informally and by structuring the planning process to focus on what the company wants to achieve.

When strategic planners depend on benchmarking to generate strategy, it can become a double-edged sword. Declining margins follow when competitors in an industry herd around a common strategy. Sustainable growth is a perennial goal, yet strategies often capture few of the intended benefits. Why do plans that look good on paper go bad when executed? Often from unforeseen and unintended secondary effect. You need to take secondary effects into account and help managers choose wiser ways to accomplish their objectives.

Your product strategy must be communicated and understood down the organisation within every department. For example can every employee relate their job to your company's strategy? Job descriptions based around the company's strategy and plans are a good way of involving people. The descriptions should indicate their duties, responsibilities and authority.

You're in big trouble if they don't understand this because you'll have a workforce that doesn't know why they're doing what they're doing. You'll miss out on the unexpected contributions they can make. Most are keen to better the results you planned. But if they don't know what you are aiming at, they won't be sure that any improvement they suggest will be in accord with the furrow you want to plough.

> If your staff do not understand your product strategy, they will not understand whether their ideas are useful or not: and they won't know why they're doing what they're doing.

Not only that, but you need to encourage contributions. The most innovative ideas tend to come from cross-function teams with capabilities in more than one area.

If everyone in the company has a simplified version of your product strategy, the ideas they generate will relate to it, rather than your having to turn ideas down that are constructive but unusable because they don't relate. You'll demotivate them when you reject suggestions they thought were helpful but weren't in line with your planned direction.

You set out relative priorities for the spend between

- products;
- technology development; and
- market development.

You set out sales targets for each new product, their contribution and proposed evolution in future business. Do this for long-term and major projects, not just short-term or incremental ones.

These goals should be the foundation of every group and department's business objectives; don't let each area create their own set of goals that are not based on the main plan.

7.10 The key factors for a successful process

The key factors for a successful process are

- Know your strategy, set it down in writing, and stick to it.
- Prioritise projects and run as few as possible in your programme.
- Get top-level support to make sure that all areas do their bit.
- Monitor and control every single stage to make sure it stays on track.
- Don't skip bits of the process 'to save time' because it won't: costly late changes will result.
- Continually evaluate how you do things.

Continuously monitor progress because, when left to their own devices, things have a habit of degrading. The resultant failures are expensive and give rise to

- lost sales because the product is wrong, over cost and late;
- the customer losing confidence in the company's decision making.

Be aware of the pitfalls that can cause the programme to go astray

- lack of top-level support for the process so that staff upstream and downstream of the technical area don't do their tasks;
- creeping compromise and specification degradation because a poor marketing specification doesn't describe what customers want;
- me-too rather than new ideas;
- so-called short cuts that give rise to chaos later.

7.11 Your company's name

Does your company's name influence your sales?

A final word on what may appear a lighter note but could be significant: does your company's name matter? How much does the market success of your products depend on it? If you really must choose or change it, what principles should you adopt? Here is some wisdom from Martin Dickson (Financial Times, 16 February 2002):

1. 'Don't dump your history. It's part of your corporate culture and your public identity. If a name is no longer appropriate, consider abbreviating it, or see if you can pull a more appropriate one out of your historical locker.

   For example *BTR* was a good substitute for *Birmingham Tyre and Rubber*. So why did it drop it in favour of *Invensys*? *Marconi*, with its historic resonance, was a good name for the telecoms rump of *GEC*, albeit tarnished by later events.
2. Choose the name of something concrete rather than abstract. *Kingfisher* is not bad. *Orange*, helped by a clever marketing campaign, has proved good. Contrast these with telecoms group *Thus*, which sounds like a fragment of a philosophical concept from *Thus Spoke Zarathustra*.
3. Try to link the name to your activities. *Bass* has been derided by traditionalists for changing its name to *Six Continents*, but it's a good name for a global hotels group. And *lastminute.com* is one of the most memorable names to emerge from the Internet.
4. Avoid names that are hard to pronounce clearly and quickly. For example *MMO*$_2$ was chosen for *BT Wireless*. Saying the four syllables out loud sounds like a Tibetan prayer chant. It quickly became known as Moo in City institutions.

5. Avoid the pretentious. *MMO_2* provides a lesson in what not to do. The name incorporates the symbol for oxygen because it is necessary for life. With its weak market share, did this apply to the company?
6. Avoid cod Latin and Greek neologisms like *Consignia*, *Innogy*, *Energis*, *Diageo* and *Invensys*. They do not conjure up any clear image of what the company does and are easy to confuse: could you tell an *Innogy* from an *Energis*? They also tend to give companies a rather temporary character – like the names given to long forgotten car models, or energy trading companies that collapse in accounting scandals (*Enron – 2001–2002*).'

There is tough competition to be top of the schlocks. *MMO_2* probably wins, just ahead of *Innogy* and *Invensys*. All were coined between 1998 and 2001.

In reality, how a company performs is far more important than the name attached to it. In 1924, Jack Cohen, a London market trader, needed to find a brand name in a hurry for a tea he wanted to start buying from a certain importer, *T.E. Stockwell*. He took *Stockwell's* initials and the first two letters of his own surname. Result: *Tesco*. Daft name? Great company.'

Chapter 8

Planning your product programme

The platform and variant master schedule

Having set out your strategy, you plan your programme to introduce the products. You set out the range you intend to develop, the variants, the common platforms and parts, and the top-level timing plan. Why do you need a formal product programme? What is product architecture and how does a well-conceived one reduce company costs?

8.1 Why it's needed

Companies need to plan a defined medium to long-term product development programme on how they will fulfil their product strategy.

Your business plan, the framework for generating income based on the revenues and costs of projected sales, will be based on your product programme. Everything will cascade from that. Products take time to develop. So your product plan is for some years ahead, perhaps five product development cycles or more. And you update it continually. The far future is less certain than the near future; so you progressively reduce risk by continual updating.

Many company management indecision about their programme absorbs half their new product's time to market. Continuous vacillation: what should the programme be, what are the product's market requirements, how might they be tackled, will it cost us too much, should we really do it at all – and so on. Most lack a formal method to give them confidence on whether to do it or not. Most don't rank projects by their risk–reward profile (explained in section 8.3.7).

8.1.1 What you do

Researching and defining the programme to introduce new and modified products needs input from all areas. Define and set out which products will be developed and when. The earlier you start your planning, the better your level of responsiveness, the less your programmes will cost and the more successful your products will be.

Two factors need definition:

- How many projects can you handle?
- Which have priority, and which will you exclude?

You examine at least the future legislative, regulatory, environmental, customers' life-style, technology and competitors' activities. You continuously research and develop an understanding of customer requirements to devise the shape of your product range: product succession, broadening or contracting the range, examining users' needs.

But it is not an end in itself. Some have built a political empire that sits on high, telling the rest of the company *We're professionals: don't meddle with us*. The result is often an island of under-informed expertise. So use staff from all areas to add their expertise in understanding and analysing the information you collect. If you don't, you may end up with anodyne product plans that will be technically ridiculed.

8.2 How do you decide what to include in your programme?

You research what potential and actual customers value to avoid planning to make a *Sinclair C5* electric car or a *Ford Edsel*. Section 9.2 deals with this aspect in detail. Most of the reasons to purchase stem from factors controlled during design and development. Customers buy because of

- product superiority;
- product uniqueness;
- after sales service;
- perceived value.

Achieving customers' needs comes only from surveying what they want, whether by post, telephone or face-to-face interviews. Section 9.2 deals with how you gather this data. And most importantly, you observe how customers use the product and what features persuade them to buy yours and not your competitor's (especially see the section 9.2 end note).

8.3 Setting out your product programme

Your programme defines what products will be developed and when. To do this, you first define what your product consists of, then set out the logic of your current and intended product architecture. The programme's purpose is not to devise the exact specification of each product. It needs a detailed study done in *Phase 2* of your product development process with wide input from all areas of the company. See section 9.2.

8.3.1 Define what you are offering the customer

First of all, exactly what is your product? Are your customers looking for just the product hardware, or a product plus unique services that only your company can provide? In particular, what support is needed after purchase? Each aspect needs to be handled in its own particular way. If you sell only the hardware, your market may be easier to extend across borders. Whereas if you sell services as well, customers like to be close to the source.

You don't develop an individual product in isolation. It's part of a logical programme, inter-linked with other products.

8.3.2 Define the logic of your product architecture

Having defined what you are offering, you set out your family of products and their hierarchy. Chart the logic of your product range on common platforms. This is shown schematically on the left. It forms a simple basis for your development programme and avoids proliferating your dictionary of component parts within your products, as well as the product architecture above them.

Product architecture:

Level 1: The product

Level 2: Platform

Level 3: System

Level 4: Building block

Level 5: Component

Each product has a logical platform structure. Some platforms may be used by more than one product. A platform is made up of systems, building blocks within systems and component parts within building blocks. Products consist of a logical hierarchy of components, building blocks and systems. This commonality reduces costs by simpler manufacture and control. And it cuts the time and cost of developing new products.

If you don't do this, your range of products and their constituents can easily run out of control and be over-costly to manage. Your staff should not be allowed to proliferate your population of components, building blocks and systems to complicate your manufacturing policy, slow development and increase the costs of control.

Malvern Instruments had this problem (see section 5.17 for more detail). They make equipment that measures particle size. In the mid-1990s they wanted to move to incremental development but the lack of commonality across their product range made it impracticable. The inventory of parts and systems was far too large. There were different microprocessors, controllers, optics and no common platforms. Typically, it took 24 months to develop a new product. The company decided to redesign and commonise its product range. Three years on, it had superseded its old product range with one based on common platforms, systems and components. Product development lead-time came down to 12 months for new products and 4–5 months for incremental models. The common modular building blocks allow fast new product turnover. This would not have been possible without first reorganising the product architecture and commonality.

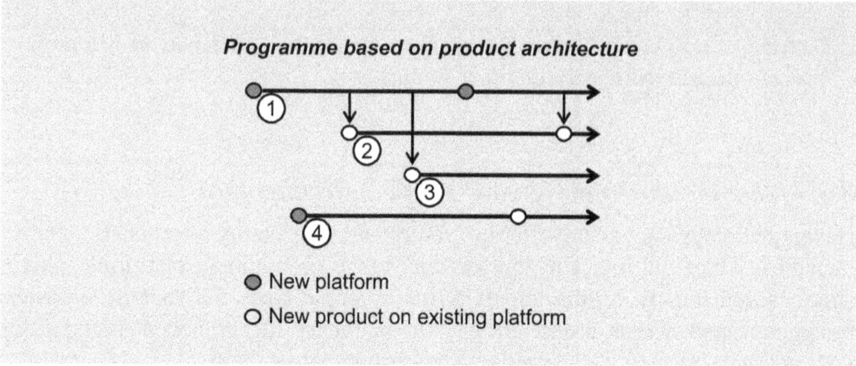

8.3.3 Relate product architecture to timing

This is shown schematically for a company that plans to use only two platforms to create four products. The chart shows their relationship and sequence of introduction. It starts by developing product 1 on its first platform. Then it develops product 4 on a second platform. Then product 2 based on its first platform; product 3 also based on its first platform. It then plans to revise its first platform; then update product 4; then update product 2 on the revised first platform. You can create a chart showing the logic of your own product range. It provides a logical foundation for your development programme.

You plan the platform strategy of your future products to match your product programme to your market areas and applications. For example

- plan variants based on a common parts or systems portfolio;
- plan your major building blocks and how products will share them;
- plan how the products will share components;
- plan your derivative products.

Armed with a clear plan of your product architecture and a programme derived from your product strategy, you will have a better chance of cutting development time and cost.

8.3.4 Continually update your product plan

External circumstances change your programme, which is continually updated. The faster you run your projects, the fewer such changes affect each one because there is less time for external circumstances to change.

8.3.5 Contingency planning

If customers like your new product better than the business plan assumed, could you produce enough? Do you have a contingency plan in place, to cope with less or more demand than you expected?

8.3.6 So how many products should you have in your programme?

We don't overload our production facility because we know it causes chaos. So we plan a margin of spare capacity for contingencies, which invariably gets used up.

> Too many projects choke your programme and, like traffic congestion, movement will be frustratingly slow.

But why do we always overload our product development capability? Most companies

- put everything into the programme because they don't realise that overloading slows everything;
- don't understand how to prioritise or if they do, don't do it ruthlessly; this is top-level task, not to be delegated to lower levels;
- do not (or don't know how to) resource-plan staff in all the functions that contribute to product development;
- many plan in outline only for the technical area; they are then held up when other areas, often process engineering, are not forewarned or don't have the capacity; or marketing won't do the early stage work: all result in a large number of costly late changes.

So the answer to the heading question is the fewer the better. There will be pressure to include every proposal. But everyone will get their project earlier if you do fewer and resource each one better.

Too many projects choke your programme's operation.

The human brain is not good at dealing with continual interruptions and frequent subject switches. (Ladies are said to be better at this than men – apparently because of their better left-brain connection.) We often assume that mental changeover between projects is instant. It isn't. Everyone who is juggling several projects makes slow progress on all.

> Typical staff utilisation, working on
>
> 1 project – 75%
> 2 projects – 85%
> 3 projects – 60%
> 4 projects – 45%
> 5 projects – 35%

When working on one, you have all its issues in your mind. To switch to another, you purge all this and load the information on the new project. It takes time, is mentally exhausting, and if you do it several times a day, you lose the ability to get things done.

All will take too long because everyone works on too many at once with resulting poor utilisation (see page 103). On two projects they can switch to the other when one is held up. But on three or more, they not only have to stop one and start one of the others; but in addition they will continually be interrupted by queries on the other projects they are not working on. It radically slows progress. Staff may look busy, but their effectiveness becomes worse and worse. And it saps morale.

8.3.7 Prioritise projects by their risk–reward profile

To limit the number in the programme, evaluate project risk against opportunity. Include only those that give the best risk–reward profile and postpone starting the rest. If a brilliant new project is suggested, have the courage to put an existing not-so-good one on hold and do the brilliant one first. This task needs to be done at director level because it sets the pace for being able to achieve sales and create wealth.

The aim is to minimise your programme to only the most profitable: overload the programme, and like a traffic jam, it chokes speed of progress.

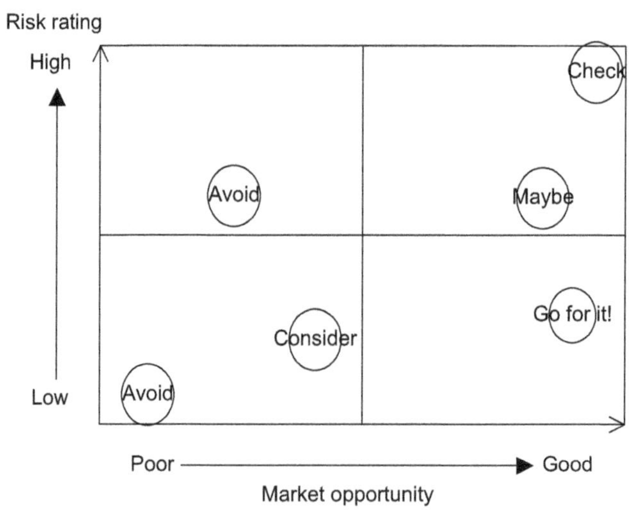

So rank projects graphically by their risk–opportunity rating (see the chart above); gut feel is not sufficient. If the market opportunity of a project is huge, but risk is unacceptably high and development cost is low, it is worth proceeding at least to *Phase 2* to confirm the risk level (as opposed to the advice 'avoid' on the left). So it is useful to add a third 'cost' dimension to the chart. *Chapter 10, part c* explains a simple effective methodology to rate risk.

Commercial opportunity (reward) can be represented in a number of ways, for example discounted added value over the product's life. But beware of grossly optimistic forecasts made by any area solely to justify starting 'their' project; make the area accountable for achieving their forecast.

8.4 Example: how TSS Limited continually update theirs

TSS Limited makes motion sensing, heading, positioning, pipe survey and cable survey systems. Through continual open communication and discussion, they make sure everyone understands their product programme and can comment and contribute. They do this through staff-away days every six months. Then every five to six weeks they have a staff lunch with a number of presentations. And the monthly managers' meeting formally reviews each project (graphically presenting targets, man-hours spent, cost, resource needed to complete each) and considers potential new ones.

They use these three meetings to identify requirements for particular products. This develops and refines their product plan continually to maintain its relevance. It's more effective than creating a plan once a year, then waiting until it's up for revision next year.

The monthly management meeting prioritises and tabulates potential new products that they evaluate in more detail on a simple chart. Some projects may be cradle-deaths; some may be viable to develop. But they know they can't do everything: they concentrate on their core business skills.

> Each potential new project is tested against company strategy. The product development requirements, staffing levels, marketing and selling issues are tabulated. They proceed to develop only the most viable within what their capacity allows.

After reviewing current items at the monthly managers' meeting, they move on to strategic issues. How is progress compared to the product plan, what are each department's targets for six months ahead? They discuss potential opportunities or threats to the strategic plan. For example if 35% of their sales come from offshore oil business, how would an oil price collapse affect their business? What could they do about it? New ideas often emerge. From one meeting came two sound ideas for new products, one for fast ferry stabilisation and one for deep-sea fishing boats. But they assess their feasibility to add them to their product plan only if they fit their product strategy.

8.5 Further detail

More detailed background to planning your product programme is given by David Allen in his excellent manual[1] on the subject. See *Chapter 12* for availability.

[1] *Developing Successful New Products – A Guide to Product Planning*, David Allen.

Chapter 9
The seven key project phases

To reduce risk, you perform a logical sequence of activities in your product development programme. Omit or ignore any, and you risk incurring ever-larger costs as the process proceeds. And to optimise time and cost, you shape each products programme individually.

The earliest stages determine the ultimate success of your products, and the costs you will incur later. You progressively reduce the unknowns step by step, and at the same time increase the value of your product. You end up with a product that has few unknowns, is what the customer values, is fit for purpose and is designed to be manufactured as simply as possible at lowest cost.

Background building block development

This is a continuous product development programme that you do in parallel with your main programmes; it may commit only 2–3% of the product development budget; its purpose is to de-bug untried or risky products, building blocks or components. It covers the development and invention of new product ideas and technologies, and the continual review of possibilities for your product strategy and product plan.

9.1 Phase 1: Pre-development

Screen the product's apparent viability: is it even worth considering from what you know internally; if the answer is yes, plan the outline resource for a project that satisfies the company's strategy with the best fit between what customers want and what you can do; check feasibility with preliminary technical and market assessments; do a fast estimate of costs and income over the product's life.

9.1.1 Introduction

As soon as an idea starts to be considered, appoint a board-level project champion (see section 5.16) to organise this initial screening.

The aim is to check, before spending any money at all, that the idea looks to be potentially viable. So before you decide to do a more detailed investigation, make sure it's not a fool's exercise by first doing a quick assessment from what you know now. This is a fast *internal* consideration.

> Many companies waste half their product lead-time vacillating over whether to start. Others start instantly and later find that the project was obviously not viable from the outset. **Phase 1** prevents both.

Be aware, before you even start, that if you economise on these early stages, you may discover that a product that appeared to be quick and cheap to design is becoming unexpectedly expensive to de-bug as it is developed. This is commonplace. You may discover, after a few months, that the project's feasibility is starting to look questionable: can it ever be made to perform reliably in service? Or, too late, you may discover that early production models don't perform as you intended, and the production process won't deliver the quality levels when you try to manufacture it in larger quantities.

All this results from inadequate input and planning in the earliest stages – foreseeing the problems that will occur later and designing in counter-measures from the start. The task of designing products is a continuously moving target – moving away from you as customers demand continual cost-down, technology support, product improvement and (your) investment. It needs as much ingenuity to devise how to design the product development process as it does to design the product itself.

9.1.2 Is there a potential project?

'Seat-of-the-pants' ideas have been known to work well; but history shows us that most fail. Why? Because the 'inventor' of the idea has not paused to think through whether people would really want to spend money to buy their idea – whether it is really technically and economically viable.

> From what you and your colleagues know, make a fast assessment of whether you have a potentially viable project or a foolish idea.

You need to answer a host of questions before you even start to commit money to the development project. But don't let that be an excuse for delay. This is a 'quick and dirty' first phase that takes hours not weeks.

You have generated a number of ideas (see section 7.6) that could have potential. But before you enter any into your product development programme, consider from what you already know whether a market exists for them, and whether they might be technically and commercially viable.

So who might buy it; in what quantities and at what price; what might it cost to manufacture; will it need to be launched by a certain date; what time and cost might it take to develop; do you have the necessary technology to develop and make it; can you afford to? (Sometimes, can you afford not to.) Is there sufficient potential margin?

9.1.3 The principles

The first thing you do is Phase 1: fast internal groundwork to test your proposal's feasibility by screening it before you even think about committing company funds. It starts off with the project champion doing a fast think through the project. Then follows a more detailed screening getting views from others. At the conclusion of this phase, you should know whether you have a potential project or a foolish idea.

It's a fast approximate justification. Does it fit your product strategy? Might there be enough potential customers and would they want to buy the product in sufficient quantity? What are the costs and could the sales revenue create an adequate return?

Right from the start, work to limit your exposure to risk. Remember that product development is strategic, not tactical. Devising a new product needs to be driven by

- what customers will buy,
- how the market can be developed, and
- what will improve the efficiency of your company.

In other words, it should be matched to your strategy. It's a carefully considered step, in response to what your planning activity has identified. It should help the company's development. It should not be done to fill a financial gap in panic, or to fulfil delusions of grandeur.

9.1.4 Your initial, fast personal screening

First make a back-of-an-envelope estimate; and if there are no really obvious flaws, get your colleagues, views, as discussed in section 9.1.6.

> Go through a fast first screening using what you know yourself.

Rapidly think your way through the whole programme for the new product idea. Go through each major step of the programme in your mind. Are there any obvious potential weaknesses in the process before you even start? This first-cut is a kind of fast back-of-the-envelope exercise that takes hours rather than days. Its purpose is to convince yourself right at the outset that your initial decision to go ahead and commit pilot resources will be justified – before you start spending money on market or technical research.

Is the proposed product part of your product programme (see scction 8.2)? If not, does it fit your master plan? If it doesn't, should you amend the plan? Can all areas that the product touches resource the project or should you delay starting because it would take resource from projects of greater priority?

First, estimate the outline project risk from what you know now. How to do this is shown later.[1] Is it sufficiently risk-free to contemplate? If satisfactory, continue. If not, consider what you might change to reduce the risk. If you can't, still continue if the potential costs and sales look promising. It's worth proceeding

[1] See *Chapter 10, part c – How you assess and manage risk.*

to a more detailed screening to examine its feasibility. You will be forewarned that you will have to reduce particular risk elements.

Second, quantify the approximate resource and income equation – roughly what resource might you need to commit, and what are the probable rewards.

Do you think the idea is technically feasible? If it looks like pie-in-the-sky, get some other views to test it. But remember, even IBM reckoned the PC was a stupid idea when it was first suggested. If it's just a fantastical wish, maybe impossible, probe its feasibility with the originator. If they can't convince you, stop now. Record it as a Board minute.

If you think it looks feasible, make an informed guess at the potential income from the product and the costs of developing, producing and launching it. What is the rough magnitude of items such as people and cash resource required; sub-contracted services; time scale to introduction; likely product cost and production scale; funding costs for the production resource and so on. Comparing that to the expected rate of generating income over the product's life arrives at the likely financial resource and return. Repeat a quick risk analysis (see *Chapter 10, part c*). If it looks interesting, proceed to the next step.

If you don't do this, you may waste time considering a project that may not be viable. For example parts of the resource could be too large to fund within the project's time scale. Or it just can't be completed within a critical time scale. Or – most serious of all – it's an unconsidered diversion from the company's longer-term strategy. You set the preliminary goal posts. Some companies don't do this. Some months later they have to cancel the project because they come to realise they can't achieve its aims. A fast first-cut throws out obvious non-starters.

9.1.5 Your second, more detailed, internal screening

To proceed with more certainty, you now require a more considered check on the project's feasibility, costs and income potential over its life. You thereby estimate the order of return you might expect. Again, this should take hours rather than weeks. This is a more considered estimate.

> Next, make a more detailed check by consulting the in-company knowledge experts.

Select the potential project manager to help do this. It will result in a better understanding and initiation into the background when you get Board approval to start the project. It gives a more sound start-up.

9.1.6 Is the product technically feasible?

As many projects fail the technical feasibility test as fail the market test. If it's a completely new product, it's a good idea to check feasibility first. Still do this even if it's only a variant or a modification to an existing project. That extra feature you want to introduce might be more risky than you think.

Are the technologies that will be vital to the product's functioning sufficiently risk-free to ensure that you won't have performance or reliability problems in service? Discuss the outline feasibility with colleagues or expert contacts. Use what they know by talking to them. Don't ask them to do research – that comes next. Write up their expert views and get them to sign them off. If you don't, they may backtrack later. I've seen this happen all too often.

Are the technologies that the product will use fully developed and evaluated? Are any areas, critical to the product's success, not fully developed? If so, can you develop them quickly to the point where they no longer pose a significant risk? If all the technologies are not risk-free, are you 99% certain that you can make them so by the launch date? Is there a fallback solution if you can't?

If they are not sufficiently predictable and there is no viable fallback position, you are going to need some magic ingredient, like an amazing potential return, to make it worth even considering taking the risk. Quantify it and set out a range of potential counter-measures; plan your insurance for the worst outcome. If its feasibility looks eminently worthwhile but definitely unproven and risky, put it into your long-term product development programme and sort out the risks off-line. (Explained later in section 9.3.2.)

9.1.7 Potential income

At the same time as estimating costs, ask your market research function to make a more accurate estimate of likely income with an outline approximation of the market and the potential for sales. Use whatever intelligence already exists within your company, not a programme of external research. That comes next. For how many years will you make the product before replacing it? Do you intend to develop it incrementally at regular intervals? Will there be variants? All this needs taking into account by recycling into the cost stage you are completing.

9.1.8 Likely costs

To launch the product, you need to estimate the cost of developing it, the costs of funding any additional manufacturing facilities and the manufacturing cost of the product.

To estimate development costs you can use past history, with an estimate of the size of team – the man-hour input – and your target lead-time to launch.

Your time to money

Target lead-time is a feature you will continually strive to reduce with every succeeding project. You should have objectives related to the kind of project, whether new or just incremental. It should be expressed in terms of time to money, not just to pre-sales launch.

> In gaining market share, speed of introducing your new technology across your product range is just as important as time to market.

Your time to market aim depends on what the market expects as well as your strategic objectives. There's no point in developing products faster than is necessary to obtain advantage. For example you don't need to develop trucks as fast as cars; or a container crane as fast as a TV set. Product complexity influences both project lead-time and man-hour content. If you don't need a short lead-time, you can spread the same number of man-hours over a longer period.

But it's not just time to market that is important to gain market share. Speed of introducing your new technology across your product range is equally important. You need to transfer your new technologies through a number of projects while the technologies are still new to the market. See *Chapter 10, part b* on managing multiple projects.

Team costs

Total costs can often be factored on man-hour input. It is usually the governing factor. Team working is vital if you want to minimise project time: so what level of team will you need? Is the product completely new, a derivative of an existing one or just a modification? Will it be a full multi-discipline team or will a smaller more limited one suffice? You'll need to estimate the type and probable size of team, as described in section 5.2. Less risky or complex projects need fewer staff and less elapsed time.

Manufacture cost

Manufacture cost may need a brief discussion with one or two strategic suppliers as well as developing your own outline in-house estimates. You also need to estimate your probable warranty and service costs, which (of course) will be low if you do everything right first time.

9.1.9 Project timetable

You now put together an outline of the potential project timetable by phase, stage and task. What phase reviews will it need? (Depends on risk level – see *Chapter 10, part c*.) Divide the project into major phases and within those into relatively short stages. This initial guess at the work schedule develops into a running document throughout the project. As the specification firms up, you can better estimate the manpower resource for each task.

When sub-dividing into tasks, note that it's difficult to monitor the progress of long-duration tasks accurately – they can slip without being visible. If a task takes more than two weeks, consider sub-dividing it. The two-week arbitrary suggestion will depend on your total time to launch. It needs to be tailored. If you go for too long without discovering that a task is off-beam, you will put too much strain on your resources to bring the programme back on target. Shorter is wiser.

9.1.10 The returns

Having outlined feasibility, costs and income, you can then estimate the product's returns over its whole life.

You thereby have an estimate of the project's costs and returns: does it look financially viable? Do you think that the estimations you've made can be justified in a real project and will the product have a sound foundation? Rework the estimate of the outline project risk (see *Chapter 10, part c*) that you started with. Is it still satisfactory and the project still potentially feasible? If it looks too risky, can you make changes to revise its risks downwards? Or can you revise costs and income – whichever is weak.

9.1.11 Phase conclusion

From asking what your colleagues know, you now have an outline of the project's financial viability, its order of risks and returns, and an idea of the type of organisation and programme you'll need. Hold your first project review meeting[2] to test whether it meets your objectives. Proceed or cancel according to the outcome. If you cancel, you might put an idea to the long-term development programme; or alter its aims to make a new viable project; or archive the information for input to later projects; or if you decide it will never be viable, bin it.

The upstream phases

Upstream processes are managed to encourage generating as many ideas as possible.

9.2 Phase 2: Researching the project to make a quantified business case

Research in detail what customers want and whether it looks technically feasible; outline a project and assess its risk and reward; define how the product will satisfy both company and customer; make a quantifiable business case; generate the product's user specification.

Idea generation
This phase is the start of the upstream processes. You use a looser management style, to encourage as many ideas as possible to be generated and evaluated – a more freewheeling approach than the pressurised, rigorous management of the downstream processes.

9.2.1 Objectives and rationale

Your business plan
The purpose of this phase is to confirm in detail whether you have a viable project. By the end of this phase, you will have researched where you are and where you want to be, will have a quantified business case and a marketing specification for the product. You will then have arrived at

- **The what**: product and market requirement, route to market, geographic spread.

[2] See *Chapter 10, part a – Programme control reviews.*

- **The how**: to resource it, in-house or out-sourced and team building.
- **The when**: time-scales, planning, management.

Quantify:

- potential risk;
- likely sales;
- project costs;
- production costs;
- investment costs;
- likely return.

This is the foundation for the next phase, where you scheme out a number of physical concepts that could satisfy the marketing specification.

In this phase, you assess and research the product and market to

- determine strategic fit and risk;
- research technical feasibility and customer requirements;
- estimate costs and potential financial return;
- make a quantifiable business case; and
- generate the product's marketing specification.

How you ensure that it sells

In manufacturing companies in the West, companies don't often research many sources of information, or measure customer attitudes, or formalise information gathering or have contact between the product development team and the customer. But these are all essential.

Without quantified information, the next phases will not have a sound foundation. You may start tilting at windmills that don't exist. This phase costs relatively little and sets the mould for the project: a product that customers will want, that you can make and that fits your strategy.

9.2.2 Staffing the project

The project champion and project manager will already have been appointed. Confirm the number of formal reviews for the programme. You'll know how to do this when you've worked your way through the chapters on the seven phases and how you manage projects. Appoint a review group to approve the project's continuation at phase reviews. And use the first project team members to assist in doing the work in this phase and to put the business case together. The earlier the team can be involved, the better their later decisions will be.

The surest ways to lose money is to introduce a product without finding out what customers might want.

9.2.3 The product

What is the product for?

You can't be sure of developing a successful product unless you find out what customers want. You may feel this is not necessary. Maybe you know of one or two brilliant products that were not researched. But was it luck? Every year thousands of products fail because manufacturers have not researched their customers.

> Don't allow your staff to become obsessed with product features and ignore the customer's problem. The product is not an artefact that is interesting in its own right. It's a tool that customers use to solve their problem – useless in isolation from the problem.

Customers buy your product because they see it as a way of solving their problem, which may be a mix of physical and emotional requirements. No perceived problem – no requirement – therefore, no sale. And this will involve its appearance and presentation, surprisingly important even for a 'functional' product; they justify its price to the customer. So

- What problem are they using it to solve?
- Would they like it to solve a problem that it doesn't at present?
- What would be an acceptable solution?
- How and why would they go about selecting their purchase?
- What price do they believe is good value?
- How much will it cost to make and sell?

Answers to these come only from surveying what customers want, whether by post, telephone or face-to-face interviews, and most importantly, by observing how a customer uses the product and what features persuade them to buy yours and not your competitor's. All dealt with in 9.2.5 below.

Don't devise your specification from a compilation of sales staff comments. Sales staff want sales results yesterday – this is tactical. Your company needs a strategic result. The two can be mutually incompatible. Therefore, investigate customers' priorities so you can relate them to the priority of the features that you put into the product. Use quality function deployment (QFD), a formalised way of doing this, described in section 11.2.

What is the extent of your product?

Develop not just the product itself but information, booklets and even customer training on items such as delivery, installation, operation, maintenance – whatever applies. Developing these is part of product development process and needs to be inserted into the time and budget of your programme. It is not an after-thought: you do it at the same time as you design the product. It is all part of your expenditure on product development that may also be important for tax and for accounts.

The appearance and presentation of the product and its associated deliverables are surprisingly important even for a 'functional' product: they justify its price to the customer.

9.2.4 *The product's feasibility: a more thorough assessment*

In *Phase 1*, you made an informed guess on whether the product would be technically feasible. You now need to check in greater depth – as exhaustively as you need to be certain – that the technologies vital for the product's correct functioning are sufficiently risk-controllable. Involve the team as well as any outside specialists so that everyone understands the issues. Those with a more general background can often point to aspects that specialists miss.

The purpose here is to see whether you think the ultimate product will be technically feasible. A product needs to specified at three levels:

1. **Concept level**: what the customer expects the product to be and to do – when, how and in what conditions (environment).
2. **Functional level**: how it is going to achieve its performance – how it works.
3. **Technical level**: the specification and constructional details.

The main objective in this phase is to first define and try to assess whether the product will be feasible at all three levels.

Visualising the product

To do this, you have to make assumptions about the configuration of the final product, before you even start. This may sound something of a tall order before you've designed it. But you've got to make a start. Some products will be easy, some won't. You'll have to imagine a fuzzy outline of the product, sufficient to consider whether it is physically possible to make it work. In many cases the new product idea will be defined well-enough in its marketing concept to make the task easier.

A relatively simple example is a new golf buggy. It doesn't need that much casting around for ideas. It can be propelled by a number of alternative motive units; its structure can be made from a number of quite different materials; its appearance could be widely different; you can add numerous features that your prospective customers might want. And you will know what markets you are going to aim at. You can make some fair assumptions about whether the end result is likely to be feasible.

At the other extreme, during the 1939–1945 war the British Ministry of Defence was bombarded with novel suggestions for weapons (see *Wheezers and Dodgers*, Neville Shute-Norway). They evaluated the technical feasibility of every idea submitted in case one might provide a major military advantage. One idea submitted was to catch the enemy aircraft in a special searchlight beam, then solidify the beam so it brought the aircraft down. Maybe you think that's funny; the originator didn't – he was in earnest. When they enquired what amazing technology would enable them to do that, he replied: 'I have no idea: it's merely a matter for your R&D.'

The scope of your feasibility investigation will lie somewhere between these two extremes.

Technology, capability and information

> Are there technologies you haven't yet used that could add significantly to the product's attractiveness?

Are they fully developed and evaluated? If not, can you develop them to the point where they will pose a controllable risk? How certain are you that you can make them reliable by the launch date? Set out the fallback options for each if you can't. And are there building blocks, which your long-term product development programme (see *Phase 3*) has now de-bugged, reliable enough to use?

The information sources your team needs to consult include suppliers of materials, specialised services and manufacturing capabilities; the skills you will need; the availability of funds for development; the customer feedback – uses and expectations for the product – and so on. The whole team becomes as well informed as possible on every aspect.

The team rounds up the investigation to agree that what the customer wants is technically feasible, using the technical and the customer research. If it can't be agreed, review what is possible to see if you can form the basis for a competitive product. If nothing can be retrieved, recommend aborting the project and archive the information.

9.2.5 Why do customers buy your product?

Existing upgrade or innovation?

> In both cases, you have to discover what customers would like a product to do, that yours or your competitors' products don't already do.

You will discover this only from researching potential customers. You may have a smart idea that you would like; but to make it viable you need to test that enough customers would like it. The idea needs to be communicated well enough to be able to do this.

Clive Sinclair visualised a compact commuting vehicle that would cost peanuts to buy, less to run, and could be repaired by the local electrical appliance shop. The shape and form of the vehicle was a closely guarded secret, his personal concept. The result was 10/10 for concept but unfortunately not even 1/10 for execution. The C5 is now an eccentric collectable.

Remember: being 'market-led' is not 'being led by your marketing department' ('s opinions).

What counts is not your beliefs or opinions but what your customer thinks is the difference between your products and your competitors'.

Why do your target customers buy other products in preference to yours? Why do they think other products are preferable; are they more innovative, better to own or use, less costly, better delivered and supported – or what? Many don't know the difference that customers see between their own and their competitors' products. Some believe they have to be the cheapest; but they still sell even though they aren't. Customers buy because of a host of factors, not just price. It's on overall performance of product, sales literature and manuals, how you sell, service, respond to their problems, reputation – and more. How do customers view your and your competitors' instruction manuals, for example?

Most users can't understand how to set up the features on their car radios because many instruction manuals are unintelligible. And even when they do understand how to, few users use all the features that have cost a lot to provide. Could the suppliers have saved themselves the money? Did they first find out whether, even if users did know how to operate them, they would value every feature? Or did they just copy what they thought other manufacturers might do?

The product's objective

A product is a way of solving a customer's problem; not an artefact that is interesting in its own right.

Your product is a tool, useless in isolation from the customer's problem. Despite this, surveys tell us that fewer than one manufacturer in twenty ever investigates their customers. Caveat: what customers want may not be what they need and vice versa. Give them what they want, but only when matched to your strategy and capability.

The best firms really understand their customers. They define the customers they want to attract. Then get behind the defined group's thinking to understand why they buy. Then investigate why other potential customers don't buy their products.

This enables them to look at the product in the same way as a potential customer. This is not how their staff look at it, from a company or personal point of view. They also try to make their products exceed customers' expectations.

Detailed market research and feedback

The attitude that *It's no use asking the customer because they don't know what they want* is usually an excuse to try to avoid doing so. Other interesting excuses range from *It's not possible to see into customers' minds*, to *We have all the information we need from a focus group*, to *We'll explain the product to the customer*. Or, from a marketing manager, *Trust me I'm a professional.*

> The aim is to make selling superfluous; to know and understand the customer so well that the product sells itself – Drucker.

You do need to research the customer, how they buy and how they use the product. You not only need to know what the product should look like, feel like and perform. But you also need to discover what influences them to buy it. There is no point in developing a brilliant product unless it also attracts customers.

For example in the 1990s Nissan attracted few customers in Europe. But when they did buy, customers were really delighted with their cars. Nissan had not tackled the vital character and features that attracted customers in the first place. But they realised and are still taking action.

You need sufficient quality as well as quantity of investigation; otherwise, you may make interesting *discoveries* later that will cause costly changes to the programme. You update the information continually throughout the subsequent phases but in such a manner that does not keep changing the specification. Your aim is to determine the best fit between what customers want and what you can do – matching it to your product strategy.

Who are your customers?

What do you understand by *customer*? Who are they? You need information not just on your immediate customer but on every level up the chain to the final user. For example your immediate customer may incorporate your product into their system and sell it to an original equipment manufacturer (OEM). The OEM might then sell it to a leasing company that leases it to the final user. In some cases, there can be a chain of buyers. For example a domestic shower unit might first be sold to a regional distributor, then to a building materials' wholesaler, then to a builder and then to the user. Each will have particular requirements; research them to make your product as attractive as possible to every part in the chain. There's no point in developing a dog food that the dog would love to eat, and the owner would love to buy, but which the store finds unattractive and therefore won't stock.

> The customer chain:
>
> - product user
> - product buying chain
> - product assembler (OEM)
> - major systems maker
> - sub-systems maker
> - piece-part maker

The data and how you collect it

This depends on your product. Typically, the starting point is researching the company's internal situation, then the competition, the sales distributors and the end customers – their quantifiable requirements, attitudes and psychology. Evaluate

- your product strategy; and
- your internal strengths and weaknesses;
- your competitors, their products, pricing, plans and strategies;
- your sales distributors' attitudes and views;
- your customers: who they are and what they want.

The fuller the picture, the better your judgement. Use the information to feed the initial QFD process matrix.

You need to survey existing customers, because your customers have experience of using your product that you don't have. Their feedback is invaluable. How they are using them – is it exactly as you thought they would? What new things would they like it to do?

> It's vital to research your current customers; but also investigate your competitors' customers to continually expand your customer base.

But also research those who are not, but could be, your customers. This will ensure you continually expand your customer base and do not miss those who are deserting you for your competitors. For example do you know why your competitors' customers don't buy your products? Research what customers want, by a combination of

- mailed responses;
- telephone interviews;
- face-to-face interviews;
- by observing how the customer uses the products;
- by observing what features persuade them to buy.

For example Scientific Generics, in researching user needs for new medical equipment, taped all the surgical procedures for 24 hours in each of 37 hospitals. They compared common issues and identified 34 major needs. This observation exercise generated six product innovations.

The scope of this research will depend on whether you sell mainly into individual large contracts (an oil platform, for example), to a limited number of original equipment or systems manufacturers, through dealers or agents, or directly to end users.

You assess four kinds of customer needs:

1. essential features that they take for granted (such as your car starting every morning – you can't sell that as a competitive benefit);
2. performance features that they value – the better it performs, the more attractive it becomes (such as doing 10 miles per gallon more than your rivals);

3. new capabilities they would value, not yet in competitors' products;
4. unobvious features they don't expect, that they really value, that delight them and that your competitors don't provide (yet).

When you study how potential customers use a product, go to the environment where they use it, observe and understand why they do what they do and see how you can provide a better solution to their problem. Also see how competitors' products fare compared to yours.

Do the same thing in the environment where they buy the product. What attracts them to buy yours? How do they make the choice?

In many cases you may be interested in winning a supply contract rather than a consumer buying a product. What aspects clinch contracts – what does the immediate customer look for? Research the customer chain better than your rival bidders.

Example

The following methodology was used by the UK 600 Group's Machine Tool Division that was determined to rectify its incomplete grasp of customers' needs. This resulted in a product that took 22% of their target markets within five months of its launch.

They steadily built up the detail through a series of linked tasks. To follow their programme you would research four areas:

1. **Internal situation**
   What is the company's plan – its product strategy, its production intentions and so on? Compile an accurate strengths, weaknesses and opportunities survey of the company. Gather internal information and data; conduct internal interviews with key people; do the same with key external people, sister companies for example.
2. **The competition**
   Understand exactly what your competitors are up to and their intentions. Research and analyse competitor trends: the technology they are using and likely future developments; the sophistication of their products compared to your own; what the customer values and what they don't; chart prices relative to specification for your and your competitors' products.

   Compile a competitor library; go to exhibitions and talk to their salesman (who can be helpfully indiscreet about their company's future model plans); get field reports from your sales force; find what your suppliers are saying about your competitors (that can be a real grapevine of information). Compile a price–specification matrix for your and your competitors' products. Set out what you think the industry's strategy is going to be.
3. **Sales distributors**
   Collect factual data by questionnaire; supplement it with sampling interviews to reinforce and refine the data; add supplementary external research. Hold regional feedback sessions at major distributors to discuss the findings; get their involvement and later on their 'buy-in' to your product programmes.

4. **Researching potential and actual customers**
   First define the target market: who could buy your and your competitors' products? Characterise the typical user in a customer profile. For example in Colchester Lathe's case, they found that 96% were small engineering firms employing fewer than 100.

Customer requirements
Use a relatively simple, inexpensive postal survey. Profile the potential customer. Use direct mail with a well-classified mailing list that can identify the customer attributes you need. Collect all kinds of quantifiable data that forms invaluable reference for your specification. Ask direct questions of fact such as what are the characteristics of their businesses and the factual, physical, quantifiable requirements of the products they buy: what do they use the product to do, what do they want of the product they use now and what would they like of future purchases; ask questions about facts, not attitude questions that may be ambiguous; if you need further detail, clarify it by supplementary telephoned questions.

Attitudes and awareness
Use telephone interviews by a researcher who appears to be neutral. What is their awareness of manufacturers in your sector and in particular how aware are they of your company; what do they think of your products and those of your competitors' – what do they perceive as the strengths and weaknesses of both.

Their psychology
A face-to-face survey: use your own staff to interview users in a formal *voice of the customer* programme. Your staff could arrange to research relevant companies in particular market areas. They design a questionnaire to extract attitude and psychology data that is difficult to obtain by postal survey or by telephone. Many of the questions are open-ended; none is leading. All are designed to make the respondents open up and explain their underlying thoughts and beliefs on the company and their competitors.

Each company is interviewed at their own premises, and the interview takes around two hours. Sales team members are not present at these interviews, to ensure impartiality by avoiding any precedents that prior personal relationships could have established. It also avoids leading questions that can unwittingly be inserted as part of a salesman's natural technique.

What are the real underlying reasons that customers buy your products rather than your competitors'; what do they really think of your company and others; what is their business outlook and their equipment buying plans over the next 2 or 3 years (use a time-span appropriate to their purchase of your kind of equipment).

You don't need to do hundreds of interviews. Experience shows that by the time you've done around 30, you will probably have surveyed 90% of the possible information.

Analysis
The data and interview answers are then grouped into common areas, analysed and a database prepared of the major points raised. Product price-against-function

charts are used to show the present market position compared with the competition, and the predicted positions at launch date. These charts create a useful debate on the merits of the proposed specification and functionality of the product.

Such data collection exercises are not costly for the results they yield. You scale the exercise to be in proportion to your company's operation. Both Clausing and Reinertsen[3] go into more extensive detail.

Organise your own staff to collect the data

If you use an outside agency to do all the work for you, they may take the most valuable knowledge – the direct contact with the information sources – away with them. Spread collecting quantifiable, factual data among members of a number of your departments to increase their knowledge.

> Your team must understand the customer: they can't get this understanding third or fourth hand.

What do customers value? How do they make their decisions? What do they want? Why do they want it? It's not just about facts; it's equally about how customers perceive a product. What affects that perception? It's almost more important to understand why the customer wants it than to know what they want.

In developing a new range of products, Eurotherm Controls Limited decided to make some significant changes. As Dr George Turnbull, then Technical Director, Controls Group, commented:

'The development teams at first sight appeared to be working efficiently and time scales were acceptable. But in terms of delivering product that was right for the market, and which met financial objectives, they were not acceptable. The general attitude was 'give us a spec, agree the time scale and we'll do it for you'. There was no questioning whether what was developed was right for the market.

So we made a significant change to the way data was gathered. We now make sure that key individuals have first-hand experience of customers and markets. They have to shape the specification according to what customers want. As a result, team members now have a commitment to the market success of the product. If the product is a market flop, they can't shelter behind someone else's specification – all members have to be convinced that what they're doing is right. The prime motivation must be user benefits. They have to devise product features that are genuinely useful to the customer.'

Therefore, do not localise the exercise to your market research area – use design, development and production members of your product team. They gain valuable insight, practical understanding and experience of what customers like or

[3] *Total Quality Development*, Don Clausing; *Managing the Design Factory*, Reinertsen.

dislike. You will get better quality answers from your staff than from a researcher who does not have your staff's extensive detail knowledge of your company's products or businesses. You'll achieve ownership, knowledgeability and motivation; your people will be more aware, better informed and will make better decisions that will save you time and bring you a more successful product.

Generating product 'newness'

If you limit your product exactly and only to what customers say they want, you may end up in a me-too game – just copying and improving rather than generating new ideas. Your competitors may end up with exactly the same ideas for their product. If you provide only the basics, customers may prefer another product that provides a more interesting solution to their problem, or has a new capability they value.

Therefore, encourage your staff to look not only at what customers say they'd like, but also at better ways of solving their problems that they may not be aware of. Your staff will know of developments in the pipeline; maybe some could improve the product to be more attractive than your competitors'. Novel ideas may emerge – like the original idea for the *Sony Walkman*.

> Some companies waste half their total product development lead-time in vacillating over the marketing specification.

Often, customers unwittingly develop a potential new product for you. They modify and adapt your own or competitors' products to suit particular needs. Use your field applications and service staff to feed back what they are doing and why.

9.2.6 The marketing specification

Ah but!

If indecisive or uninformed pressure from any area prolongs your project by vacillating over the marketing specification, you won't generate income soon enough. Through delaying the launch, the manufacture cost and the project cost will escalate. Typical examples of indecision are

- Pressure from listening to too many voices (often in the sales and service areas) can proliferate into product-range overkill. Creating a vast product range bewilders customers and can tempt you to stray dangerously away from your core competencies.
- Pressure to set a target cost too low can ruin customer appeal. History shows that customers buy on a combination of specification, performance and price; they rarely buy the cheapest available product irrespective of specification – they buy on perceived value within their budgeted price range. If your price is way below this, many buyers may not even select yours for evaluation because they think it can't possibly fulfil their needs at such a low price.

- High-level individuals within the company can press their prejudices on the team contrary to the evidence of the data collected.

The marketing specification – areas to consider:

Primary and secondary uses

Nature of user/operator

Product features required:

- function/performance
- appearance
- user interface
- dimensional constraints
- graphics/corporate identity

Patents and licences

Service requirements:

- operating duty-cycle
- operating environment
- transport/delivery/installation
- security and safety
- potential product liability
- storage/preservation/protection
- reliability/maintenance/disposal
- manuals for installation/operation
- operator/maintenance training plan

Key standards, regulations and safety requirements

Target market
Target sales/production rate
Related company products' replacement/obsolescence
Competitors' products
Variants/developments
Price/cost strategy
Product life span
Programme/project/launch timing
Project cost
Plant/tooling amortisation
Sales, distribution, packaging
Manufacturing facilities and methods:
in-house, sub-contracted, licensed

Look out for these and have counter-measures ready.

Don't give away features that customers don't want

Make sure that potential customers don't think they're paying for features they don't want. Use the first stage of QFD (see section 11.2) to ensure the features you specify are only those that customers value. This not only helps generate ideas but also avoids wasting money on 'good ideas' that customers don't value, which you end up giving away free – a recipe for reduced margins and fewer sales.

Setting out the specification

Use the parameters of the marketing specification to feed the first QFD matrix, prioritising customer needs and wants.

Having collected sufficient data to make you aware of the spectrum of facts and views, you are now in a position to devise a firm marketing specification for the product that reflects customers' perceptions rather than those of sales agents or distributors.

Why the reference to customers' perceptions of your products and company? Because it is expensive to change customers' perceptions: it can require a costly advertising and public relations programme over a long period. On the other hand, dealers can be educated and their attitudes influenced more quickly and at less cost. And you can use your new product development programme to catalyse internal company changes.

Prepare a draft market requirement specification from evaluating a trade-off between

- features and performance that customers want;
- what is feasible for the company;
- target manufacture cost including funding any new investment;
- required launch date; and
- the cost of the development project.

To do this

- assimilate the marketing information;
- use relevant information on competitive products;
- establish the intended product status in the market;
- include likely variants;
- set out the performance targets that will meet customers' needs;
- identify probable critical problem areas.

Typical areas you will want to consider in your marketing specification are shown on the left. The specification is a description of what the customer wants of the product, not how you intend to achieve it. And it is a control document against which you judge later achievement.

You may want to devise sub-specifications for particular systems of the product; the same rules apply. It's a description of what the customer wants of the product system, not how you intend to achieve it.

Variants and new technology

Refer to your product master programme (see section 8.2). If variants are needed for different applications, expanded versions or adaptations, or for application in different markets, include this requirement in your marketing specification. You will save time and resource if you plan for it now. It will cost you dearly if you wait until you are well down the road to developing the product, often to the point of its production release, before you decide to create variants.

You should aim at a specification that can be modified and evolved to form the springboard for a series of products, using the same basic platform technology. (See *Chapter 8.3.*) Your investment will yield a higher return over a longer time frame. It also helps you generate a consistent brand identity. You can even derive quite different products from a common base.

For example Black and Decker developed their first heat gun from one of their electric drills. Only when it had achieved sufficient market success did they develop a unique one with further variants. This considerably reduced the initial risk of generating a new family of tools.

Carrying over new product features to further projects

You may generate some new concepts or technologies in developing this new product. You will generate far more sales if you can carry new unique features and building blocks over to other products at the same time, whether existing or new products. Spreading new unique features across multiple projects needs to be done quickly before they become stale. Achieving apparent newness of products has been found to expand sales. Relatively few companies take advantage of such concurrent transfer of features or technology. Speed is key to the results.

But to get innovative new designs, you first have to develop your new technologies and do it continuously as background product development. And this continuous background development is better done by specialist functional (or system) departments rather than by product teams. You also need good planning to ensure that the new, shared building blocks arrive on time for your cluster of new projects. You will need to manage suppliers' programmes as a part of your in-house programme. These are all factors that you need to plan into you programmes.

If you expand your product range in this way, do not expand a single project to accommodate these new variants. You won't be able to manage it adequately. Instead, use a single base project and generate separate parallel projects, transferring the technology across as they run side by side. And don't wait to complete one project before starting the next. If you do, the teams may not generate common building blocks between variants; they may make unwise compromises in their designs and consume too much time in co-ordination work. That will lead to loss of product uniqueness.

9.2.7 Example: how TSS (UK) Limited assesses project viability

TSS (UK) Limited gives a good example of the kind of attention a small company gives to these Phase 1 and 2 processes. They employ about 50, and develop motion sensing, heading, positioning, pipe survey and cable survey systems. Says Robert Manson, their CEO:

> Before we spend any time and money on developing a product, we evaluate its potential. We use a detailed two-stage process to evaluate the risks and opportunities.

'We have to be market-led. We talk to customers every day, analyse new product opportunities and investigate new markets. But before we spend time and money developing a product, we always evaluate its potential. We use a detailed two-stage process to evaluate the risks and opportunities before seeking approval from the board. The assessment will depend on the type of product and where it fits, either within one of our existing markets or a new market. It applies whether the product is an in-house idea or from a customer. The evaluation is the first part of our new product development process.

Phase 1: The senior management team examines the idea to see whether it could be viable. Initial filtering must be good because Phase 2 takes resource away from other work. We need to be really sure that it's worth proceeding.

The Phase 1 outline specification for the Phase 2 evaluation is never really adequate. That's why it's so important for the senior management group to analyse the key issues thoroughly first. We used to use a checklist to formalise this process but we found that people just ticked points off without sufficient consideration. So we made them sign off the key issues. That ensures proper analysis and commitment and the results stick.

Phase 2: If the answer to Phase 1 is positive, we move to getting solid market feedback to convince ourselves it's viable. We discuss and examine it in detail with potential customers, using a non-disclosure agreement. If that looks positive, we review the return on investment. The team evaluates and estimates

- the market and likely sales;
- the time and resource to develop it;
- costs and cash flow;
- how feasible is it to make the product – especially any special processes.

We rely on effective teamwork to achieve the level of confidence we need. From this, we know the potential returns and costs in detail. It tells us whether the project is worth pursuing or not. Most importantly, it avoids committing resources on non-viable products. From there, we can produce a business plan with a good degree of certainty.'

9.2.8 The business plan

You can now generate a business plan for the new product:

- the product opportunity;
- the potential market – define exactly who will buy it;

- define possible variants;
- sales targets;
- an outline of the product's characteristics – the marketing specification;
- an outline project plan of tasks, resource and time;
- project risk;
- capital and revenue budgets;
- likely costs, margins and cash flow.

This requires financial analysis, re-assessment of the project's risk and a business assessment. You examine the potential competitive advantage that you will achieve.

Define the criteria that will be used to assess project success: market share, sales volume, overall return, margin, customer satisfaction, break-even point, product performance relative to target, quality and reliability in service, project budget, team satisfaction? Decide which you will use. Ensure these are in the phase review criteria.

As part of the business plan detail you will also consider items such as

- an estimate what you think your future position will be, your reputation, the size of the market, the expected demand and your target position and share;
- estimate the impact that your competition will have on this;
- create a SWOT analysis of current and competing products to guide the development;
- then define the positional aims for the product in its expected markets: product variants, product costs, production rates and so on;
- specify the launch timing requirement, specific windows and so on;
- estimate the time to ramp up to full production; therefore, when does production need to start;
- specify the total funding needed for the project, revenue and capital;
- estimate approximate unit production costs, margins, cash flow; although difficult at this stage you may have targets you want to meet that can be set as project targets; accuracy will be refined as the project proceeds;
- build up the programme timing and cost in outline; allow for incremental revisions;
- assess your match with production and other in-house capabilities;
- assess what supplier partnerships could add competitive advantage to the programme;
- define distribution channels, production and service facilities, and new facilities needed;
- set out a phase review timetable and review-team/method/dates;
- define the training and planning workshops to help implement the programme;
- do a 'what if we don't do it' evaluation (if you don't do this, you may be in danger of losing market share).

Typical business plan content:

- summary and recommendations
- description of the proposed product's performance requirements
- rationale and strategic aims
- product-market aims, market segmentation
- external developments, competitive products
- product positioning, impact on existing products
- target sales territories and territory variants
- product requirements and technology
- legislative provisions
- product cycle plan
- product development programme timing, launch, critical dates
- manufacturing programme, location, suppliers
- financial aims/targets
- potential income: volume, price, cost
- project cost: capex, revenue
- resource: manpower, equipment, cash
- financial viability: sales and cost targets
- project risk versus reward analysis, fit with product programme
- conclusion

And when you are embarking on developing products, don't be afraid of devising a business plan that looks over simplistic. Almost anything is better than not having one at all.

In the early 1990s, Infrared Engineering set out to revamp its range of measuring instruments. As Dr Roger Edgar, then engineering director, remarked: 'As far as the business plan for the project was concerned, it was so childishly simple that I hesitated to reveal it:

The basis was that we sold about 60 units a year of the old design at about $10,000 with a net margin of about 20%, yielding a net profit of around $130,000 a year. With a modernised design, we could increase the margin, probably to over 40%, at the same sales price. Given that the new machine would be better than the old, we could either increase the price and the margin further, while holding unit sales constant, or perhaps increase sales whilst holding the price constant. Either way, our profit was likely to be usefully greater than a cautious $250,000. The probable scale of the development would be about four man-years, and since product development is highly labour intensive, for our purposes a good enough cost estimate was $75,000 a man-year, totalling $300,000.

And even if the figures didn't look half as good, we would probably have gone ahead because the existing design was so long in the tooth that if we did nothing, its sales would soon have dwindled.'

Not all business plans are like this. An evolutionary replacement of an existing product in a niche, near monopoly, market is probably about as simple and low risk as you can find. But for most developments, beware of business plans in which the outcome would be changed if any one of the numbers changed by (say) 10%.

9.2.9 Phase conclusion

You now have a business plan for the product, and its marketing specification.

Re-check its risk profile (see *Chapter 10, part c*), which should have reduced. If it hasn't, can you amend the project's aims to achieve a satisfactory risk level? This might be done by altering the commercial, technical or market aims, depending where the unacceptable risk lies.

Hold a phase review to determine whether you've completed all the tasks necessary to proceed to the next phase.

A final note

> Much attention is focused on speeding up product development time and adding new features to products. This may miss an increasingly important issue of how consumers access and use these products.

As Dan Jones and Jim Womack note, just finding out what customers want in the product alone may not be sufficient. You also have to understand how to support the product after they've bought it. They comment:

While speed and accuracy in developing a product may be necessary conditions for survival, they may not be sufficient. Being closer to sophisticated consumers should give local manufacturers an edge in understanding their current and future needs. But much of the attention is focused on speeding up product development time and adding new features to products. This may miss an increasingly important issue of how consumers access and use these products. At the same time many companies are losing direct touch with consumers as they outsource customer support and help lines, as well as production.

You need a simple method to define value from the consumers' perspective, the first principle of Lean Thinking. What might lean consumption look like as a complement to lean production and lean supply chains? The key is to recognise that buying a product is not an isolated transaction but a series of steps just like production: it is a consumption process. And its purpose is not the product itself, but the use of the product, together with the relevant services and knowledge, to solve the buyer's problem.

This can be on a personal or business level. In our professional lives, we are both consumers of products and services from others and providers of products and services to our customers. There is a consumption process, mirrored by a provision process, mediated by a series of interactions between them – telephone calls, exchange of information, delivery of products and so on. As with production, we can list the consumption steps and record the time they all take, carefully differentiating the few steps that help consumers solve their problems from the rest. We can also note how helpful or frustrating the interactions with the provider were and whether they were necessary or not.

Learning to see the consumption processes of the different types of consumers that use your products is becoming a key company skill. To recognise how badly designed and dysfunctional both these processes are just reflect on your recent experiences as a consumer, whether consuming on your own behalf or in paid time on behalf of your business.

It would not take a rocket scientist to redesign the provision process to save time and cost for both providers and customers. It would be a brilliant place to start thinking about new ways of solving customers' problems.

Food for thought when you are doing your Phase 2 work.

9.3 Phase 3: Generating concepts

This is where you insert innovation. Devise several fundamentally different concepts that meet the user specification; model, simulate and test prototypes of new building blocks to evaluate their risk and feasibility; scheme how to make it; model the concepts' appearance.

Few books devote much space to this phase, although it is arguably the most important one after analysing what the customer wants. In this phase you generate a number of different concepts and evaluate each against the marketing specification from Phase 1. It's a freewheeling ideas-generating process. The next phase optimises your concept: there, you analyse what you want to incorporate to make an even better concept. The important thing is first to generate a number of good, different concepts, containing sound new ideas.

> This is the phase that will mould your future market success. It's the one place in your process that deliberately gives you time to devise new or innovative ideas.

Generating concepts sets the foundation for the success of your product. It's the one place in your process that deliberately gives you time to devise new or innovative ideas. These are essential for your future, to avoid continual me-too products or ones that are obvious re-dressed previous ones. In such situations, potential customers have a habit of remaining potential rather than becoming actual. (That is subtly different from incremental innovation – see section 11.5.)

It is also the one and only near-zero cost opportunity you have to feed good ideas, new ways of shaping the product, novel features and smarter ways of being able to make it into your new product. Go beyond this phase and you'll have missed the opportunity. Later changes will cost you an order of magnitude more: so changes two or three phases later will cost you 100 or 1,000 times as much.

> Not even a magic wand will transform a lousy concept into a brilliant product. Or 'great Kaizen can't make up for lousy product design'.

Concepts are solutions that meet the needs of your *Phase 2* marketing specification. Concepts are the 'how' that satisfy the 'what'. You may be devising changes to particular systems rather than to the entire product. You consider the concept of each individual system.

9.3.1 Why bother to do this?

Before you even start, you may have a feeling for what a good solution would be. So why bother to generate more different ideas? You might believe it would waste valuable time: or that you could continue indefinitely and suggest concepts forever more. So set a time limit.

If you don't consider several ways of doing it, you are unlikely to find the concept that gives anywhere near the best solution. In other words, one that contains all the features the customer wants at the lowest cost and that you have the capability to organise competently. Some big failures have resulted from over-confident, *got to get on with it* cultures of not doing this.

Single solutions are not recipes for success. Many want to stop when they've found the first possible solution. Resist the temptation; if you go gung-ho for the first idea that comes into your head, it's likely to be far from optimum. But if you consider several, you are likely to follow the 80/20 rule. In only 20% of the theoretical time to reach a 100% solution, you will get an 80% one. It's extremely cost- effective.

The benefits of avoiding late changes

Some of the most cost-effective product developers are using the concept phases to cut 30–40% out of the time and cost of developing their next generation of products. If you skimp this concept optimisation phase, you will incur a large number of drawing changes later, possibly even post-launch because of your unhappy customers' discoveries.

> Effective use of the concept phases can cut 30–40% out of the time and cost of developing your next generation of products.

The most common reasons for changes to post-production-release drawings are

- **Unwelcome 'discoveries'**: concerns raised by marketing, manufacturing or service when they see the product because the concept did not incorporate all their requirements and misinterpreted others; this can account for the largest proportion of all; you will have dealt with this in Phase 2 by having everyone from all areas input their knowledge into the marketing specification.

- **Simple mistakes** that were not discovered, usually through inadequate care by individuals, for example from inadequate checking for internal tolerance clashes.
- **Revisions to reduce costs**: usually because it's too complex and insufficient resource and brain-power has been given to generating an elegantly simple concept; it's not easy, but you can always make it simpler and easier to make (one company mistakenly committed to investing millions in 'advanced manufacturing systems' because they quickly devised a vastly over-complex product).
- **Specification changes** from insufficient initial planning; inadequate research into what customers really value; or poor market intelligence on what new developments are imminent; or because it took far too long and the market changed; or perhaps a specified material could not be purchased.
- **Changes** resulting from the product not attaining its performance targets, often because insufficient risk analysis was performed when the concept was formulated or because features were incorporated that were not fully proven during the concept stage; and from lack of product robustness, dealt with below.

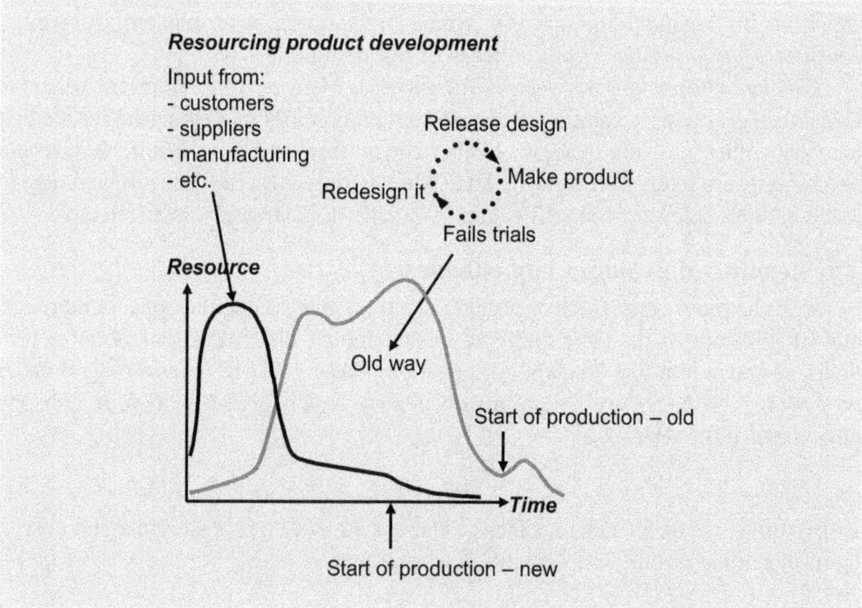

Of these reasons, the first can typically trigger about a third of the changes, and the remainder about a sixth each. The effect that high numbers of changes have on your programme's time and cost is truly staggering: it can double both.

9.3.2 Who does it

Expand the project team further from all functions, led by the team leader appointed for *Phase 2*. *'All functions'* means just that: every area the product touches on its way to the customer. Leave any out, and you'll make expensive 'discoveries' later.

Project resourcing – the old way wastes money

There is a second good reason to put adequate resource into concept generation: it saves you untold project time and cost.

Traditionally, companies have concentrated most of their development effort on detailing the product and testing it. The early phases, especially the concept phase, are usually skimped because of the desire to *get on with the real task*. This is fundamentally the wrong way to go about it. Insufficient early effort pushes most of your changes to after production release, where each costs you 1,000 times more than here.

Your cost and man-hour input will correlate with the number of design notes, changes and so on, that you issue. Fewer changes cut development cost. Aim to achieve most of your changes before the completion of the concept phases.

This will fundamentally change the shape of your resource curve. Typically, it shows a peak during detail design, one during testing and making it manufacturable, and another when your customers find that changes are needed to make the product reliable. Instead, it should have a single peak at the confluence of the concept and detail design stages.

Suppliers

Which suppliers are strategically essential to produce key components and subsystems for you? You may be buying up to 70% of the value of your product from suppliers as components and sub-systems. Bring the most important in at the concept stage to work as a part of your development team. If larger than you, they could contribute significant technology – an advantage you could not afford on your own. Such collaborations are especially common in the Far East and result in smaller companies developing their market share much faster.

> Use strategic suppliers' capabilities to extend your own.

Use their capability and resource as an extension of your own. They can reduce the floor space and brain-power that you need to employ. And don't spend yours on developing areas that do not contribute to the uniqueness of your products. Let a specialist supplier do it with you.

While a good supplier's collaboration can be an extra development resource, a poor supplier can retard your programme and increase your costs. Are their product development capabilities as good as yours? Should you be helping them to improve? It's worthwhile helping the supplier so they operate in the manner you require. Sometimes you may have to help them solve their internal problems because you may have no option but to use that particular supplier.

9.3.3 Customer perception

A potential buyer may not always appreciate the internal elegance of a product design. However technically fascinating your new product idea may be to you, the customer may not be interested by its finer – or any – of its hidden technical points. They are unlikely to be concerned by the number of components in an enclosure, or whether new technology has been used to achieve the working characteristics.

> Customer perception governs what the customer will pay, not what it costs you to make.

But they will be impressed by how the product looks, how it feels in their hands, how well it works, how easy it is to operate and maintain, its reliability and by its price. The only reason they buy is because they think it's the best value means of solving their problem. They buy what it can do for them. You will have researched what the customer thinks is important. This needs recalling at every stage of developing ideas.

> Customers buy what the product can do for them. They want the best value solution to their problem.

> Thomas Edison realised that it's less risky to offer a solution that customers want than to persuade them you've solved a problem they didn't know they had.

Appearance is disproportionately important

The product has to look as though it's worth the price you are going to ask. There is often great benefit from involving product design specialists from the outset. Customers pay more attention, consciously or subconsciously, to appearance than is given credit. Even the aesthetics of functional and capital products, such as a machine tool, have a great influence on their sales success – as one UK company discovered to its surprise.

Some European companies, notably Italian and German, have paid great attention to this for decades. It is highly cost-effective to use industrial design input and advice right from the start of concept definition when you are creating a range of alternatives. Its cost can be insignificant compared to the return from the additional sales you will generate.

The appearance of a product is absolutely intrinsic to its function and appeal. It's not something you can add at the last moment to make it look attractive. You need to involve the concept of aesthetic design (the discipline of industrial design) right from the concept stage. It can introduce new ideas of how to configure the product and has often challenged previous precepts on the product's architecture to great effect.

For example ICG Limited developed an entirely new concept colour graphics scanner. Their industrial design function questioned the traditional appearance that was governed by the architecture of having a horizontal scanning drum, like a lathe. Everyone used horizontal drums – it was never questioned – like (almost) every car has four wheels. They realised they could develop a vertical one that could cut the floor space needed by 80% – a real plus for the customer. They could make its appearance more attractive and the novel solution increased their sales by 70% in the first year. In the first year they also recovered their whole product development investment. It transformed the company's financial fortunes. If they had just looked at the engineering aspects of the product they would never have made the step change.

9.3.4 The principles of the concept phases

Introduction

Having defined your product ideas and researched your potential customers, this phase considers different ways in which you can achieve what they want. Define the products and variants you want to develop, and when. Exactly what configuration will the product be?

And second, do sufficient testing to make sure that each building block, new or old, is reliable enough to take forward.

This phase is a diverging process, where you generate creative ideas and turn them into possibilities. In the next phase, you converge the ideas into a single preferred concept.

Here you investigate different ways of satisfying the marketing specification as well as variants. It's a creative phase, not to be trammelled or steam-rollered. You may generate some ideas that seem really stupid. But these often catalyse novel but sound ideas. Don't cramp your peoples' style or deride their ideas – they will create a winner for you. All the ideas will be analysed in due course and the unusable ones discarded.

Defining concepts

Concepts need to be sufficiently well defined to make sure they satisfy your aims. To recap, defining a product requires information at three levels:

1. **Concept level**: what the customer expects the product to be and to do.
2. **Functional level**: how it is going to achieve its performance – how it works.
3. **Technical level**: the specification and constructional details.

Failure to define any of these levels in sufficient detail will cost you money. You will encounter problems ('discoveries') later in the programme.

Concept content

The principle of concepts is to investigate new and better ways:

- performance enhancements through materials, new technology principles, new developments of components;
- novelty and innovation – novel ideas for building blocks or new ways of combining them;
- appearance to excite the user;
- new technology to permit the product's user to do things easier, better or faster;
- using physical principles that are different from similar products (e.g. cleaning clothes by ultrasonics rather than shock water movement);
- simplified construction to make the product more reliable, less costly, simpler to manufacture.

Creating the concepts

List the customer's performance requirements in a matrix (see section 11.2, QFD) and see how each requirement can be achieved by different means. Use the marketing specification as the control. Some of the requirements will develop naturally into sub-ideas. The team should put forward and discuss different ideas to create concepts for each product sub-function (system): deep probing of ideas to match the marketing specification requirements.

> Have nothing that you do not know to be useful or believe to be beautiful.
> – William Morris.

By taking a combination of the best of the ideas, create a range of innovative concept schemes for possible products. Every member can contribute valuable detail at this stage to make the product more viable. Develop a number of sketch models to outline the concept (pencil is faster at reiteration and development than computer-aided drafting (CAD)).

> The purpose of concepts is to generate new and better ways.

When you have generated several, work them up into feasible concept designs and put aside the ones that aren't. Use the production engineering and production team members to scheme the practicality and cost of potential production routes.

You may need to try out parts of designs with breadboards, simulations and so on, to check whether the ideas are sound enough to develop further.

Consider every potential aspect of the product's performance and construction, considering every detail and proving out everything that may pose any risk. You simulate the performance of the product, preferably by computer model validated against test models: very time- and cost-effective. This results in an accurate understanding of every aspect of its projected service performance and life, of its intended production processes and their capabilities, and of its likely customer impact.

Typical activities, depending on the nature of the product, are

- understand the key or unique features needed to satisfy buying motives, derived from the marketing specification;
- develop concept schemes by progressively adding engineering detail into design layouts; define the packaging of the internal detail; generate preliminary parts lists; simulate performance by computer modelling;
- produce on-screen renderings and surfaced models;
- produce full-scale mock-ups and models for viewing and critique (use rapid prototyping where appropriate);
- test new modules; develop test rigs to verify the performance of parts and systems; test trial concepts to customer duty-cycles to simulate real use; seek customers' and suppliers' input and comment.

For more background on generating concepts, Pahl and Beitz[4] examine the technical detail of this phase at greater length.

9.3.5 How you generate ideas

Concept ideas require widespread input. Specialist staff can generate them, for example within a function or system department. Outside input can be contributed, such as from searches of library research, licences or patents. And your internal multi-function team can generate ideas where not everyone is expert at everything – more a broad general view. Experience shows that a more generalist multi-function approach will generate more novel concepts than a narrower single department approach. But the single department can usually refine the detail more effectively and originate interesting new technological solutions to particular detail problems. You need both. Re-cycling between the two groups produces good results.

> Ideas are so fragile – so easy to miss because they can be so quiet, or to snuff out an idea. The inquisitiveness and the willingness to try is so important for design, for developing those tentative fragile ideas into a real product.
> – Sir Jonathan Ive, Apple.

Brainstorming

You will have collected data on requirements from the customer and from key internal areas, customers, suppliers, service, promotion, distributors and competitors; and data on market trends. Use brainstorming (see section 11.6) to catalyse ideas that customers would not have thought possible. Reach a consensus, and

[4] *Engineering Design,* Pahl and Beitz.

agree on an amalgamation of ideas. Make sure that the result meets your requirement specification.

Separate generating ideas from criticising them.

Conceiving an idea – being creative – is difficult. Whereas anyone can criticise an idea once it has been generated. Therefore, separate the process of generating ideas from commenting on them, just as you would in a brainstorming session (see section 11.6). If necessary, do it at different times. Encourage quantity of ideas. If generating them becomes a problem, consult references on ideas-creation and lateral thinking. You may want to devise solutions right back to fundamentals; TRIZ is useful for that.

But be sure that when concepts are generated, contributors are prepared to sign for their commitment:

In one complex project, we schemed out a new concept from discussing in detail what was feasible with suppliers and all the engineering areas that would contribute separate systems. Outline schemes were discussed in depth with every area – what was possible, what was feasible, what was not. We went for the safe option every time and turned the concept into general arrangement drawings. We then went back to every area to check items of information. Without exception, when they looked at the new drawings, every single area back-tracked on what they had previously said was a safe concept. They now realised that it was really going to be made and they would be responsible for their part! The detail of the concept had to be completely re designed and the project lost 5 months.

Learn from past mistakes

Most companies lose the opportunity to reduce their development costs because they don't learn from their own history and repeat mistakes. They retard their year-on-year progress. They prolong their development time because issues appear – problems discovered later on when testing the concepts – that occurred in previous programmes. They all have to be rectified with late-stage design iterations and changes. Cost and delay.

Use history drawings to avoid repeating design errors.

Some companies generate 'history' arrangement drawings that encapsulate and annotate information from previous product concepts. They use these to put all the information from past experience into their new product concepts. What was successful; what was not; what must you avoid; what enhanced them. It's all on one drawing. You complement that with a slim manual on recommended standard practice (mainly on what worked well in the past): why did prior practice succeed or fail, with examples.

Evaluating ideas

There is a big difference between critical constructive evaluation and destructive objection. Raising objections and destructive criticism kills all ideas, bad and good. Teach your team to criticise constructively; use ideas that may not be feasible to catalyse generating better ones.

> Don't let destructive personality traits negate potentially productive ideas.

Be aware that you may need to deal with a number of destructive personality traits that will demotivate or irritate the rest of the team and waste everyone's time. Shingo identifies three:

1. Type 1 will raise objections based on convincing-sounding but wrong theory; they often have no real experience of what actually happens. They will make wrong assumptions and promote misunderstanding. They are often authoritative and persuasive. Don't allow them to carry the argument on false premise.
2. Type 2 will tell you that every idea is not possible and won't work, without properly considering the problem. They will probably claim to have tried them all before without being to substantiate any of it. They can waste endless time arguing their case. Otherwise known as an 'Ah but'-er, we used to reckon that one of these had the capability to neutralise the constructive efforts of five team members. They often took a pride in their technique. Usually it was an excuse to do nothing – born out of laziness.
3. Type 3 will try to destroy the whole process by cynicism; they've never needed to do this before and it's a facile idea, a waste of their valuable time (their time is not nearly as valuable as your project).

Be prepared: don't let your objectives be torpedoed by ignorance or prejudice. Enjoy the challenge: contemplate and consider beforehand what imaginative counter-measures you can employ. Educate to change attitude.

9.3.6 Testing and costing concepts

Finding failure modes as early as possible is an important challenge. You will not understand the performance of a system just from investigating the performance of its component parts. They interact and produce unexpected results. It's not the number of components that governs behaviour but the number of interfaces between them. Don't think that just because a part is reliable in product A it will be reliable in product B. The configuration will be different and may lead to different failures.

Try out the concepts in prototype form: check for any interaction problems within the product's suggested building blocks. Do sufficient testing to ascertain that the proposed solutions are reliable enough to take forward. And have a fallback solution to ensure that the concept you put forward is feasible when all alternatives are exhausted.

There will be several categories of building block in your concept ideas:

1. Those that have been used before or are known to be sufficiently risk-free to accept without trial; you can take these forward.
2. New or modified ideas that are clearly feasible but that have not been used before and where feasibility is almost but not quite totally certain; these need prototypes to be built and tested to confirm feasibility.
3. Those that have problems where the solutions are clearly feasible but which need more time and resources to resolve than are possible within the time or budget of the project.
4. Problems where you cannot see the solution at present, but where you know that it can be solved in the end.
5. Those with problems that you don't know can be solved at all.

Categories 3–5 will upset the time and cost of your programme. Send them to your long-term background development programme. Use them as concepts in later projects when they are thoroughly de-bugged.

And if you are a smaller company that does not resource blue sky internal research, you might consider category 5 problems as research contracts for a local higher education or contract research establishment.

> Tests that don't produce failures don't produce much information either.

The value of test failures

Whatever you do, you must test at this point. The earlier you test, the less it will cost you. And make sure that your tests produce failures; they must be severe enough to do so. Otherwise you will not derive any information from the test results. But use representative testing, or the failures may be caused by duty cycles not related to actual use (obvious but often missed).

For example a UK court criticised a crash helmet for giving inadequate protection solely because the internal protective layers could be removed by hand from the inside of the helmet! This hardly represented its mode of use in an accident, where realistic tests showed it to be one of the safest.

Just knowing that it does not fail is not enough. You need to know why it could fail so you can build in measures to counter abuse or freak conditions in use. If you don't obtain the information at this point, you may have to repeat tests on complete products when they fail in the field. The cost will then be orders of magnitude higher. Don't fear failures. When you understand the possible failure modes, the modifications needed to avoid them often cost very little. This is revisited in the next phase on testing your chosen optimum concept. Reinertsen gives further explanation and guidance on why test failures are useful in *Managing the Design Factory*.

Costing the concepts

Cost the proposed concepts in outline detail with the production engineering and purchasing team members. Outline how each will be made in relation to your make or buy policy. Use the 80/20 rule to quantify the most significant costs. Break the scheme down into assemblies and components and estimate costs, most accurately for the most costly parts, more approximately for the low value items:

- generate the bill of materials;
- divide it into sub-systems;
- list every likely and known item and component within each sub-system;
- if thought prudent, add an unspecified contingency parts list to each sub-system;
- cost estimate each sub-system;
- use three categories of parts: cost the most expensive 20% accurately; the next 20% roughly; and the long tail by estimate or history;
- put the costs in three categories: known exactly; thought to be within 15%; and those where the error may be greater than 15%; continually update the table;
- it's important to develop the cost accuracy of the most costly 20% as quickly as possible.

> Cost the highest cost items accurately and the tail of least value items approximately.

You may also want to analyse the cost-effectiveness of the features. Both Pugh and Clausing give detail information on functional cost analysis (see section 11.8). You set out a matrix that shows the cost of each part against product function. The proportion of cost that contributes to a particular product function can then be totalled. You see the cost of achieving each function, to understand where you need to make improvements.

Customer–ownership costs

What it costs the customer merits further comment.

In managing the programme, considered in *Chapter 10*, you generate a model to guide you on how you control the project to your priorities. But this is from the point of view of the company. You also need to know what the customer economic model will be – what will the product cost them to own? That also requires a model.

> Model what it will cost the customer to own the new product.

What is economically important to the customer? Few customers ever work it out fully. What will the product cost not just to buy, but to install, commission, use and perhaps dispose of? The customer can incur costs of

- acquisition;
- installation;
- commissioning;
- floor space, services provision;

- running cost;
- maintenance and repair;
- disposal.

By charting these for each concept, you will understand the relative importance of each to the user. Express them graphically in terms easily understood by anyone who needs to be persuaded. And it will depend on the type of use the customer puts it to. You may need more than one model. But having set out the basis, this is quickly done. Reinertsen explains such models in more detail.

It's important to do this because users rarely do it for themselves. It can be a useful tool to promote the product to them.

9.3.7 The background development programme

Which technology building blocks can now be used and included in your concepts? And which ideas or technology developments not used before are now mature enough for risk-free use? If there is doubt, perform a risk analysis to justify proceeding with them or not. This applies even to products such as clothing where using an untried new material can lead to expensive problems when customers discover what you haven't.

> A continuous background product development programme is essential to de-bug new ideas and technologies off-line.

Do not move into the next step with any technology or idea where you have the slightest doubt about its reliability. Develop it off-line, as a continuous background development programme, not as part of your current mainstream programme.

Putting the development of unproven ideas or components into your main time-line is a recipe for time and budget over-runs, or even worse, catastrophic warranty costs and loss of reputation. If you attempt to sort out untried ideas in your main product programme, you will inevitably be late and over-budget. You cannot control how long it may take to make untried ideas reliable.

Only when the ideas are de-bugged, fully understood and reliable, then, and only then, do you move them across into the concept stage of your next product development programme. Incorporate them only when proven, when the risk level is understood. And move a member of the department that developed it into the product team, to speed the transfer of knowledge. Knowledge is usually transferred fastest by people. Don't rely on arms-length data – it slows down applying it.

To do this, you run a separate programme on developing new, untried ideas; it could be only 2% of your product development expenditure. Do not mix this with your main development time-line – keep it separate or, if a small company, have one or two people devote part of their time to it. If you are a product-team–based organisation, your product development (engineering, technical, etc.) department is the best place to run it. They will have the best repository of specialist knowledge.

In your background programme, you make, test and evaluate trial products or prototypes to confirm the feasibility of untested areas that have not hitherto been used in production. This applies equally to bought-in parts and sub-assemblies. Get your suppliers to do their component development and testing ahead of your programme. Make sure they don't include unproven concepts in their main development timelines too. If they do, they will delay your programme. This should be a key discussion point when you are agreeing the tasks and time for their collaboration.

9.3.8 Presenting and reviewing concepts

Presentation

Don't attempt to start evaluating the concepts until you run out of ideas for generating more, or you run out of time. But do allocate enough time to achieve a good group of concepts. (A group of one is not acceptable.)

> It's essential to explain the concepts by whatever method will make sure that everyone understands them; otherwise they can't give you constructive comment.

Scheme the ideas by whatever method will make sure that everyone will understand them. This can involve computer modelling, physical modelling, breadboard electronics, simple functional demonstrators, verbal and written explanation – whatever combination produces the best communication. You should scheme production methods to interact with the designs and improve them. Each should be costed in outline as part of the process.

At this stage none should be discarded because someone or some group considers the concept poor. Leave that until later. Even 'poor' concept ideas often contain useful feedstock to add to others.

Provide a period to incubate the ideas, measuring and assessing them against your objectives, to result in the proven concepts that you will take forward. This also gives you the time to revisit project definition, function, cost, customer base, route to market and manufacture. Time spent in getting it right here will save you untold time and cost in later stages.

Reviewing the concepts

To help keep projects on track, you must continually refer to the marketing control specification generated at the outset. If you don't, your product may suffer specification creep, an insidious virus that infiltrates even the tightest programmes. It causes drift away from its original targets. A concept will have four types of features that are very noticeable to the customer[5] :

1. **Essentials:** their absence kills sales (*customer: Yes, but where's the …?*).

[5] see *The Lean Toolbox*, Bicheno, 2nd edition, p. 170, the Kano model.

2. **Improved performance items** (*customer: it's better than the xyz.*).
3. **Innovative items:** unique selling points that provide competitive advantage (*customer: Now that's a really unexpected clever idea!*).
4. **Turn-offs:** features that add cost but which the customer does not value (*engineer: Now what could we put into that space?*); they inhibit sales because customers don't want to pay for items they don't value.

Category 1: make sure that these are all present.

Category 2: you develop these all the time – the reason for new products.

Category 3: the more innovative items you can incorporate, where risk-free, the greater the advantage.

Category 4: avoid at all costs! Beware of engineers whose creativity does not benefit the customer, and loses sight of the product's intent. In the end, over-elaboration kills the product's sales.

Consider each separate concept against a number of criteria, for example

- is it feasible;
- is it attractive;
- what will it cost the customer;
- is it sufficiently risk-free to be able to be taken forward;
- will it meet the requirements of the marketing specification?

Use peer group technical reviews. Bring people with particular experience and expertise to review the design. They may spot what the design team missed. Review each concept in the light of strategic importance. They need filtering; deciding whether to be filed for the future, discarded or accepted for further development.

9.3.9 Summary

You will need to complete a number of tasks (some or all):

- review the requirement that the product must satisfy;
- generate several different concepts;
- freeze your list of possible concepts to be evaluated;
- construct models to visualise each concept including colour and graphics;
- make sure everyone has seen, understood and commented on them;
- develop outline concept design layouts;
- make prototype parts to evaluate new building blocks;
- develop test rigs for performance verification;
- assess the manufacturing methods and estimate product costs;
- assess what will be made in-house or bought-out.

From this you should have a group of competing feasible concepts that you can take forward to the next phase. Review and update objectives, time scale, budget and product status. Review and update the project planning in line with the selected concepts.

9.3.10 Protecting ideas

Whatever route is taken along the development path, you may need to protect your new ideas. Protection can be expensive; but defending your patent can be orders of magnitude more so. Can you afford this? But if you have a landmark invention you could lose a lifetime's revenue if your idea is copied. It's a risk balance.

> If you protect ideas, can you afford to defend them?

If you choose not to protect your designs, you will have to move faster to keep ahead of the competition. Make sure yours gets to market first. And just before your competitor introduces a copy, introduce Mark 2 that puts theirs in the shade.

9.3.11 Phase conclusion

This phase will have generated a number of concepts. It will also have resolved any problems caused by the interaction of components or systems within the new individual building blocks of proposed concepts.

Re-check its risk profile (see *Chapter 10, part c*), which should have reduced again.

Hold the design review meeting. Generate your own checklist to ensure all has been considered and all tasks are complete. For example: Have you made enough progress with enough good concepts to proceed? Have you evaluated each against the marketing specification? Has everyone seen and commented on models? Have new building blocks been tested and either accepted or put to the long-term development programme? Or is there an alternative product that could satisfy most of the aims at far less cost?

9.4 Phase 4: Optimising and trialling the concept

Analyse each concept's strengths and weaknesses and generate an optimum final concept; scheme out its production processes; model, simulate and test the product's engineering and performance to de-bug new building block interface problems; model its appearance; ensure that each selected building block will attain its performance requirements.

This phase generates the best concept – the one that will generate the greatest success, that has the least risk and that best satisfies the marketing specification and your strategy. This phase also identifies interaction problems between the combination of building blocks you intend to use.

9.4.1 Developing the optimum concept

Selecting the best concept for company and customer is not lightly done. There is a methodology you can apply to get a result far superior to just accepting one of your *Phase 3* concepts.

Make sure your customers like it and that it works.

The aim is to generate a concept with the combination of features that best fits the market requirement and which at the same time best matches your overall strategy and capabilities. In principle, you evaluate the features of each alternative concept against the marketing specification; which best meet the specification; can costs be made low enough to be acceptable; are there details that render parts of some concepts more practicable? Can you combine the best features into a single 'best' concept?

Pugh concept selection methodology

An excellent formal technique for doing so was devised over years of successful application by the late Professor Stuart Pugh of Strathclyde University. (See section 11.7.) Its great advantage is that having developed a number of alternative concepts, the technique stimulates the team to devise better new variants. The one finally selected is invariably better than those previously devised. It almost always results in a new concept that did not exist previously.

You can develop a better concept than those already devised.

You can use the technique to develop concepts for system architecture, subsystems or individual components. The design configuration and production methods are considered together.

The method selects the criteria that are important to the customer and company, and forms an evaluation matrix. It then compares how well each concept satisfies the requirements compared to a chosen datum. From evaluating the ratings and enhancing the positive features, new ideas produce an even better concept.

You will have a number of concepts that will have been checked against criteria of customer needs and company strategy. The marketing specification criteria are entered as column headers in a matrix. Each concept is a row in the matrix. Selecting one concept as the datum, each concept is scored as *plus*, *minus* or *the same* against the datum for each of the criteria.

Strive for perfection in everything you do. Take the best that exists and make it better. When it does not exist, design it. – Sir Henry Royce.

The scoring is not straightforward. Initial perceptions change and the resulting discussion develops new insight. New hybrid concepts that were not previously considered will emerge. At the conclusion of the discussions, each concept will have a score. The best is taken forward. But awarding scores is only part of the methodology: equally important is the insight the technique generates to devise new concepts.

From the initial exercise, many of the original concepts will be ruled out and new ones added. The team will now have a better understanding of the issues. The exercise is therefore repeated, using a new datum. This results in selecting a dominant concept that will usually be a new one developed from running the technique. The one chosen will undoubtedly be better than the ones originally brought into the exercise.

The technique avoids rework later in the programme when your team comes to realise that a far better way exists. Clausing[6] describes the technique in more detail with examples.

You will also want to repeat your function costing analysis as a final check, described in section 'Concept definition – costing the concepts', and in section 11.8.

9.4.2 Concept design and proving

Generate full-design schemes, using computer simulation, rapid prototyping of components – whatever is appropriate for your products.

Learn from your mistakes

Most companies repeat mistakes and don't learn from their own history. By doing so they waste time and cost and retard their year-on-year progress. Recap on the comments on this aspect from the previous phase.

Simulation and testing

Get to the point where, by the end of this phase, everything is tested, simulated and evaluated. So that when you generate the launch specification and when in the next phase you build the first production products with trial parts, you will just be confirming what you've already ascertained. You simulate building your first optimum concept prototypes in the computer wherever possible to avoid the costs of expensive hardware. It is far less costly to computer-model parts of the product than it is to build lots of prototype parts, test them, then make design changes and keep repeating the process until you get it right. The hardware and testing costs, and the time needed, is what traditionally has taken so much time in developing a product. This is not the way to do it, as was explained at the outset of the concept phases.

Instead, you generate an initial model in the computer; then make and test a trial system to validate the computer model. Do your iterations in the computer: the model can then be developed to somewhere near a practical optimum before you make and re-test final hardware. The most critical systems can be modelled and when the final product is made, all the unknowns will have been investigated and you confirm your predictions with relatively little confirmation testing. The computing power and software to do this has halved in price about every 18 months over the past decade, and is now affordable by even quite small companies.

[6] *Total Quality Development*, Don Clausing.

By concentrating effort at this stage, you save significant expensive hardware costs before a series of expensive tests are done on the complete product, or costly facilities and tooling are irrevocably committed.

It's far cheaper and quicker to use simulations than to have to repeat design iterations with a myriad of drawing changes because prototype after prototype product produces failure after failure in simulated duty-cycle testing. Or alternatively, you've designed it so butch that you give redundant and costly parts and material away to your customers. You need to ensure that your design is somewhere near optimum.

It's very expensive to try to optimise your design by reliance on testing hardware alone. In some cases it's impossible. Computer simulation will generate a new understanding of how your systems work and how changes affect their performance.

An example of real-time simulation is that done by Martin-Baker, world leader in the design and manufacture of aircraft ejection seats. These are very costly, complex products. This example may be more than what many smaller companies need but it nonetheless shows how increasing complexity can be handled. It will not be long before small companies will easily afford such facilities. As the late Ken Yates, their programme manager commented:

> Simulation achieves a fast payback in meeting schedules, reducing risk and controlling cost. That's why we do it: it saves us levels of cost and time and permits us to achieve optimum designs that would otherwise be impossible.

We use extensive computer modelling in our systems engineering disciplines to model the performance capabilities of the seat. Martin-Baker has developed a range of in-house simulation tools for the dynamics of seat ejection. These new electronic tools permit us to assess the seat propulsion phase of ejection, the stability of the seat across the entire performance envelope, and its full trajectory characteristics. These include clearance capabilities from aircraft on the ground and in-flight. It accurately simulates the variation of ejection seat aerodynamics and other performance features.

We use thermo-mechanical models to help design the performance of seat pyrotechnics. As we gain greater confidence in using these techniques, inherently expensive practical testing becomes less necessary. These new techniques are undoubtedly reducing risk. When we have to make an expensive (several hundred thousand dollars) high-speed ejection test, we are now far more confident that the seat will operate as we predict. We increase confidence still further by refining the computer model with data from these tests.

In the mid-1990s we started to use computational fluid dynamics. We expect this relatively young discipline to play an important role in the future design of our seat systems. It is already showing tremendous benefits in the aerodynamic design of the seat. Among its advantages is that it mathematically replicates, as opposed to simulates, aerodynamic flow conditions around the ejection seat and occupant throughout the escape flight trajectory. Solutions are for full-sized seat and occupant as opposed to scale models, and for actual velocities and atmospheric conditions. So there is no concern about matching parameters, such as Mach or Reynolds numbers, since escape flight conditions are matched exactly.

These new techniques are now demonstrating significant advantages in meeting schedules and cost constraints while reducing risk. This is a benefit to both the company and to the customer.

Dealing with failures

Testing is where you learn and generate information to control cost.

> Failures are not a distress purchase: they generate valuable information.

Use test regimes severe enough to cause failures. Reinertsen[7] suggests that you obtain the most information when the failure rate is around 50%. Otherwise you may not understand the causes of potential failures. This will save you orders of magnitude of cost compared to letting your customers discover them for you. You might be able to proof the product against such eventualities for a relatively small cost when done now.

When you find problems caused by modules interacting destructively with others, the problem can be so difficult to fix that it can disrupt your programme timing. Many such cases can be fixed by analysing whether abandoning the new module will make a significant difference to the customer's perception of the product. Often it does not and you can use an existing module and put the problem child across to your long-term development programme to de-bug off-line.

You always need a fallback solution for when such problems occur. If you commit to using an unknown with no fallback, be prepared to lose a lot of profit through a severely delayed launch. This is not a tenable position. You must devise your fallback solution at the outset of planning your optimum concept. Revisit the comments on testing in section 9.3.6.

A key principle of lean product development

Avoid detailing the product so you have a build–test–fail–redesign cycle. Simulate the product, validate it with test models, optimise it in the computer and only then build a final optimised test product. Try to simulate the properties of the final off-tools parts. And do it in this phase, not after you've detailed it for production release. By doing it as early as possible, you will achieve the performance, life and reliability of all the technologies and parts proposed at minimum cost.

> You don't detail the product then run a build–test–fail–redesign cycle.

[7] *Managing the Design Factory*, Reinertsen.

Following detail design, it should then be necessary only to build a small batch of early production products, not prototypes, to confirm finally what you have established. You build and evaluate any prototypes now, in this phase. So during this final part of the concept phase, simulate real use. Be certain that every area will perform as required and will be reliable, including having simulated their production feasibility.

Do not go into the launch specification phase with concepts that are poorly tested and understood. If you do, your first trial builds after detail design will result in 'discoveries' that cause large numbers of drawing changes. You will need a second or even third revised trial-build programme to do what should have been done in this concept design stage. That is a recipe for burning money and time.

Variants

What variants are you going to make provision for? Use common platforms, not fundamentally different ones, based on different unique building blocks, each of which requires their own production facility.

> Don't generate fundamentally different products as your variants; use common architecture.

Design the concept so that you create the variants in your product range in the last possible stages of assembly. This will make it far easier to manufacture variants, and will reduce your component variety. It will also simplify how you manage your production.

It will come from using a modular approach to the design of your product.

An example is UK Triumph Motorcycles, which produces four different engines based on a common cylinder module. Fitted to a common base frame, the power unit, chassis and bodywork variations are fitted as final operations to produce around 100 visibly different model specifications.

Materials

Consider the specification of the materials, finishes, bulk and surface treatment for every part. Assess the availability of the materials specified. This sounds trite, but 'ideal' materials are sometimes specified that can't be supplied within the time – if ever (e.g. most of the steels listed in national standards are not available off-the-shelf anywhere).

There are many relatively new processes and surface treatments that can transform the durability of individual parts, and which are highly cost-effective; for example using titanium nitride plasma coating on fasteners to make them proof against seizure from use at high temperature.

> Using the input from a competent materials engineer can generate huge cost savings.

At this point there is no substitute for the input of a competent materials engineer – it can save the cost of employing them many-fold, but few companies ever use them. Specifying the optimum materials can make radical improvements to overall cost, both of the product and of the production stock of raw materials. If you can't afford to employ the expertise directly, in the United Kingdom, the Institute of Materials (UK) runs The Materials Information Service that has local staff who can advise on materials selection. Also, the Cambridge Materials Selector[8] is a useful tool; every company should have one. It sets out how to select the best family of materials and within that, which particular one. If you've never used it, you could achieve major improvements and cost savings. Visit their website, www.grantadesign.com, to discover more.

Beta site trialling and approval testing

Use optimised prototypes for user trials with collaborative customers. you may also be able to start approval testing with some authorities where they will accept pre-production product.

9.4.3 Achieving design robustness

Having chosen the concept, check that it can be made sufficiently robust. It must perform as your customers want in all conditions they are likely to impose on it. Without this stage, when your product is in production, many examples will almost certainly not achieve the performance you intended for them. This will be for three principal reasons:

1. it can perform differently depending on how and in what conditions the customer uses it; robust designs perform well in all conditions because tolerances will have been set for the parameters that are key to its performance in all conditions.
2. variations in how the product is made will cause its performance to vary and some products will be unacceptable to the customer; a robust design will have identified the level of control needed in the critical processes to avoid this.
3. products will deteriorate in use, for example from wear and tear; you can predetermine by design the extent to which they do so in order to limit performance degradation in service.

> The objective of a robust design is to improve customer satisfaction and reduce cost.

The objective is to improve customer satisfaction, and to reduce the development time and cost needed to optimise the product. Optimising robustness as early as possible in the product development process reduces later rework and therefore total time and cost. See in section 11.16.

[8] *Materials Selection in Mechanical Design,* M.F. Ashby.

This is not the same as a planning to run a zero-defect production policy. You could produce all products with zero defects to the tolerance levels you have set. But without going through this robustness exercise, you will accord equal importance to every tolerance. And as a result, your tolerance levels will be wrong – some will be too wide, most will be too narrow. So the product will vary widely in its performance even when everything is manufactured within the tolerances you set.

9.4.4 Phase conclusion

You now have a preferred concept that is proven in most aspects of its performance and requirements. You will have controlled it against the marketing specification developed in *Phase 2*.

Re-check its risk profile (see *Chapter 10, part c*), which should have reduced again.

> The key to doing it this way is that you reduce the risk of failure from all aspects by making sure of your ground before you commit major funds. And you save time because making changes is far faster in the concept stages than thereafter.

The next step is to detail the precise specification of the product you will manufacture. Hold your phase review to approve continuation.

9.4.5 The advantages of well-resourced concept work

If you follow the method described here, you will have been through several iterations of the product's features to arrive at an optimum concept. But you will not have committed significant resource on production facilities or tooling. You will have decided on the detail of your product, tried it out and made numerous changes without incurring major cost.

The reason you do it this way is to delay setting the concept in concrete until the last possible moment. It gives you the maximum ability to change your mind at the minimum cost. Until now you have been able to make as many changes as you want, and to incorporate any new information that may have arisen since your *Phase 2* work, at small cost.

You have involved staff from every area of the business to make sure that they have been able to input every aspect of information and requirement into the product. And you are now a major part of the way through your development programme. You can now complete in the knowledge that you should not be in for major surprises, and most important, with very little change that will cost you money. Your ultimate aim should be to have no changes to any drawing after production release.

9.4.6 Toyota 'set-based' development

Toyota goes about *Phases 3 and 4* differently. To summarise their process would take a lengthy paper. But the essence is as follows.

Toyota concentrate on off-line acquisition of knowledge through test and experiment to build up a universal knowledge base. Little new technology is developed as part of a specific vehicle programme.

Instead of developing several separate concepts as in *Phase 3*, they concentrate on developing different sets of solutions (i.e. set-based) for each building block or system. Theoretically, this produces a large number of possible combinations. For example eight systems, each with four possible concepts, can be combined in 1,680 different ways.

They explore these possible sub-system solutions in parallel, using analysis, design rules and experimental test work to characterise them on A3s, explained in *The Process* below. The Chief Engineer uses this analysis to motivate, liaise and trade between system development teams and suppliers to achieve the requirements of the marketing specification (i.e. QFD-driven). He has responsibility for the success and profitability of the product but no formal authority over any of the hundreds or thousands of individuals working on its development. He leads by knowing more, seeing further and asking better questions than any individual or single department. He progressively narrows the set of solutions.

The Chief Engineer may typically have 30 years experience of the detail and of the complete car. All through his career his expertise will have been developed through mentoring. For typical responsibilities, see section 5.16.

They delay making the final decision until the last possible moment but, once selected, it is not changed. That is always the key to a fast process.

The marketing specification is used to arbitrate conflicting constraints. Lean production starts with design. They examine all aspects of the car's production, designing to meet the target manufacture cost. They set the manufacturing area the target of zero engineering changes after production release. Having resolved all the key issues at the concept stage, they then focus on a tightly controlled scheduled issue of the production-release drawings.

The culture

Their culture is based on the rigorous thought process by which everyone approaches development: individual experts are not responsible for technical excellence; everyone is trained to do it.

They concentrate on off-line acquisition of knowledge through test and experiment to build up a universal knowledge base that expands continuously. Little new technology is developed as part of any specific vehicle programme. Toyota thereby avoids serious quality issues and ramps up production quickly. The culture is people-centred to develop technical expertise. To that end, all engineers spend time in factories and showrooms. They rely on people and avoid locking technical data into computer software or specific techniques.

Knowledge development – the key

Toyota engineers concentrate on building, recording and codifying their stock of knowledge on A3s (see *The Process* below), then draw on it for product design. The system creates knowledge that is stored and able to be used on any project; it is not just project-based. And the system maintains its quality and ability to be used on any product.

They look at the problem, analyse the root cause, model potential solutions and design experiments to prove the solution. The knowledge is organised and recorded on 'A3' papers, so it can be codified, and compared. Findings are summarised and trade-off curves separate the feasible from the infeasible. They use this data to design concept alternatives. They

- understand the underlying technical principles of the design;
- record and capture the knowledge in easily usable form (the A3 papers);
- design using that knowledge;
- expand the knowledge into reusable sets of solutions.

This one-page standard A3 format is the heart of the system. Every engineer learns how to read and prepare them. It enables rapid clear communication and helps clarify thinking.

The process

Toyota culture eschews administrative project management in favour of organisational learning that is focused on continuous technical knowledge development and improvement. It is aimed at satisfying customer needs. Whereas in the West, our standard processes focus on task compliance rather than learning efficiency. Their design reviews focus on the knowledge necessary to make the design decision.

> Toyota culture focuses on continuous technical knowledge development and improvement rather than administrative project management. Whereas in the West, our standard processes focus on task compliance rather than learning efficiency.

Toyota's system usually takes half the time and less than half the resource of a similar Western project. Trade-offs in set-based engineering select a combination of proven technology to meet cost targets and customer needs. Their project management is focused on technical targets rather than on compliance with a process or budgetary constraints (although they do run their projects to a universal process template). It results in only minor manufacturing problems (compared to normal Western systemic ones), total schedule adherence (rather than almost none in the West) and a concurrent joint development ethos rather than adviser expert role/DFM. The focus is on achieving visible results rather than executing tasks.

Every manager and engineer learns the common scientific approach to problem solving from the day they join the company. It is staggering how much attention Toyota continues to place on problem solving and on improving the basic stability in every activity.

They assume that even the best processes will be subject to continual and frequent interruptions. They track progress in real-time and design lean response processes so that plans are met every day. They also actively filter out noise in order to create as much stability as possible.

Conclusion

Toyota's process is strongly Toyota culture–based and difficult to replicate in the West for a plethora of reasons. So unless you can reproduce Toyota's culture and systems in their entirety, you will usually get better results from using the phase and task structure set out here. Off-line knowledge acquisition and multiple concept generation will start to parallel Toyota's activities and provide the detail that you need to achieve similar results. When that is well matured, you may be able to enhance it further.

The upstream/downstream interface

9.5 Phase 5: The launch specification

Create the detailed launch specification that meets the user specification, together with its materials and production processes.

9.5.1 Fixing the product launch specification

To reiterate, defining a product requires information at three levels:

1. **Concept level**: what the customer expects the product to be and to do.
2. **Functional level**: how it is going to achieve its performance – how it works.
3. **Technical level**: the specification and constructional details.

Failure to define any of these levels in sufficient detail will cost you money; it will cause problems at subsequent stages of the development programme because you will discover that areas have to be re designed to rectify detail that has been inadequately specified.

You have considered in detail the first two levels, and have arrived at your optimised concept, with the major uncertainties tried and tested. You now set out the detailed technical-level specification for the physical product you want to manufacture.

This is the launch specification.

The marketing specification describes how the customer wants the product to perform. Whereas the launch specification describes the physical attributes that achieve the customer's performance requirements.

> The launch specification is the detailed technical specification of the product you will manufacture.

The launch specification is a document that describes the product in detail based on the concept work you have just completed.

All the uncertainties will have been examined; all the alternatives resolved into a single best concept. It will form a definitive instruction on what has to be accomplished by the team in detail-design and confirmation testing for the new product. There should be little that is not tied down. The objective is a document that will provide the team with a defined product they can design in detail as fast as possible without specification change.

Produce it as a document that everyone in the company, and in your strategically important suppliers, can access. Set it out so there can be no misinterpretation by anyone. Use it as the control document for refining the downstream programme, from detailing to full production.

From the specification:

- split the product into major assemblies with a cost target for each;
- prepare to trade costs between assemblies to reduce total cost;
- set a target for a major reduction in parts-count compared to previous products;
- check that you will use only proven technology.

Confirm the market planning objectives:

- what is the marketing strategy;
- exactly who are the customers;
- what is the market potential;
- align production methodology with sales objectives.

Review and update the product objectives, product status, project timetable, budget and resource requirement.

The downstream processes

You now have a product launch specification. Freeze it: your objective is now to get it detailed, validation tested and into production as fast as possible. If like the Roman Emperor Nero you fiddle around the edges rather than tackling the main problem, and you make minor changes to your launch specification, you are admitting you have failed to complete the preceding phases.

> Downstream processes are managed tightly, to time and cost.

Senior management's role in these phases is to provide resource, and to remove short-term pressures and obstacles to progress. The objective must be a better product, better performance, lower cost, customer benefits and meeting the company's strategic goals. You need tight management of the process to achieve tight time and cost targets. The longer it takes, the more the process costs and the more sales you lose.

9.6 Phase 6: Detail design, plant installation and commissioning

Detail the design, generate manufacturing, sales and service information; put the means of production in place and commission it.

As detailed drawings are issued, final plant and tooling can be acquired. The longest lead items are detailed first. By the time all the drawings (or instructions) are issued, most of the plant can be in place and commissioned. Much can already have been ordered before this phase or, in collaboration with key suppliers, work already started and some tooling.

9.6.1 The principal features of detailed design

The aim here is to produce final detail instructions for production and for the purchase of materials and components. By the end of the phase you will have locked in 90% of your future manufacture costs, both revenue and capital. You will also have determined your company's added-value performance and productivity, the cost of investment, of production scheduling and of production consumables and services – in fact most of your company's cost base.

In the previous phases you've done most of the hard work in reducing risk. The concept of the product should be exactly what the customer will want, and be a good fit to your and your suppliers' capabilities to manufacture and supply. You now turn the concept into detail to produce the manufacturing instruction. Although detailing is more akin to editing your final concept, clever detailing needs to be creative and requires great intellectual input to produce a good result. Don't underestimate it.

Background on this phase is to be found in Pahl and Beitz[9] and in Pugh.

No innovation: control it to the programme target

It's too late for innovation; all that should have been done in the preceding phases. Don't start altering or 'improving' the specification either; you'll miss the launch date and raise the costs. The launch specification is already frozen. If you have to unfreeze it, you are admitting that you haven't completed previous phases' work thoroughly enough.

Put all the good ideas that emerge to the concept stage of the next update.

[9] *Engineering Design*, Pahl and Beitz; *Total Design*, Stuart Pugh.

New ideas are bound to arise during this phase. But unless they are a part of the detailing process, store all the good ideas that emerge for the concept stage of the next update. On no account mix this design phase with some fundamental research or new principle or technology. This is not the time or place. It should have been completed and proved long ago in previous phases.

Poor detailing can produce major disasters

First and foremost, incorporate prior lessons (see the next section).

The performance, reliability, life, attractiveness and every other attribute of the product is designed into it. You can't create it later by manufacturing or testing it into the product. You will have taken this into account in progressing through the previous phases. But inattention to detail design can still lose you more market share than most other factors, through ineffective and unimaginative detailing.

Use the most experienced, qualified and professional design skills that you can afford. It's the most cost-effective thing your company will ever do. Employing intellectually and possibly more expensive, more able design staff can save the company disproportionate amounts of money later on. Despite this, manufacturing industry generally tends not to use the best engineering practices and experience; you have a huge opportunity to gain competitive advantage by doing so.

Poor detailing can be catastrophic. It can cause costs from needing subsequent changes and 'refinements' from errors. You should aim at fewer than 10% of your total drawing changes occurring during this and later stages. It will save you an unimaginable amount of time and money. If this is not your current performance, you can achieve it by careful detail work in preceding stages, and by making it one of the targets for your development programme.

Incorporate lessons from previous products

Be aware that most companies – even large household names – don't learn from their own history and frequently repeat design mistakes. By doing so they increase their development costs and retard the way they introduce better products. They prolong their development time. Because they repeat previous errors, late development issues appear from problems discovered in testing concept prototypes and early production examples. They all have to be rectified with design iterations and drawing changes that are unnecessary. They simply don't learn from experience; they waste countless, unquantified time and money they would find hardly credible if they monitored it. In some instances it has doubled their product development costs.

*A famous retired aircraft designer was always welcome to visit his old company. He was encouraged to walk around the design office to see the new projects – he was always fascinated by engineering detail. One day, he stopped at the layout of a new tail plane. 'That's interesting', he said, 'Why are you detailing that area with (this that and the other) feature?' Having listened attentively to the explanation, he said 'Do you know, that's exactly the reasoning I used when I designed the tail plane of the *** 30 years ago.' He then explained the errors he had made 30 years ago that were about to be repeated.*

Use history drawings to capture past lessons and avoid repeating mistakes.

Don't invent sundials or re-invent the wheel. Incorporate all the detailed information from past experience into your new product. You can do this by creating 'history' detail drawings and specifications that encapsulate and annotate information from previous products and projects. You create a scheme of a particular area of your product and annotate it all over with notes. Why did you detail a particular area in that way; for example why is a particular dimension critical. What was successful; what was not; what must you avoid; what enhanced previous products; why was it done that way.

At the start of the detailing process for each new product, the design engineer pulls out the history drawing from the previous product and assimilates all the notes. At the end of the design and proving phase, they add the new experience gained to the drawing, for the next project to use. You may need to complement that with a manual on recommended practice: why did prior practice succeed or fail, with examples.

Standardise on design methods and specifications

It's important to standardise not only on components and building blocks but also on the methods you use to design their functionality.

Use a consistent approach – set standard design methods to solve specific types of problem that occur routinely. Don't allow your staff to use unchecked or ad hoc untried methodologies. Record the approach taken and the reasoning behind it, for future design problem work and for post-mortems. This could apply to calculating performance, stress, response of machines, system stability and so on. Update the process notes as better techniques or tools appear.

Also set out guidance on product design criteria that are not available from textbooks, and specify those that are. This could be for items such as permissible loadings and safety factors for your particular product, your preferred materials range, their application and so on. And standardise how you present the design information, both in format and completeness, on specifications, formulations, process instructions, drawings, data sheets – whatever you commonly use.

Limits, tolerances, finishes, materials, welding practice, fasteners, adhesive bonding methods and materials, plating, anti-friction and anti-scuffing treatments, heat treatment: all these need agreed standards not just internally but to ensure correct performance from purchases. You don't usually need to create your own specifications; just have instruction sheets on which commercial ones to select from. The list needs continual review to see if it can be simplified. If you do create special company specifications, this usually results in higher costs; they need to be generated sparingly and used only when absolutely vital to performance.

Standardise materials too

A small competition carmaker decided to standardise on a minimal variety of materials specifications and stock sizes. Some were far better than was needed for certain parts.

But they gained real benefit from having a vastly reduced variety of raw materials to stock and control. Their annual raw material stock turnover increased from 2 to 14. They cut their raw materials inventory to only 15% of what it had previously been.

> Although specifying more costly materials may appear to be overkill for some parts, even if only a few parts use it, you can save huge inventory costs to buy only one more costly material in larger quantity.

The alternative is to stock small quantities of a profusion of lesser materials that are on the face of it cheaper than the overkill specification. You probably end up with ten times as many individual stocked combinations of material and size, occupying more space, taking more time to control and lowering the stock turnover. This is significant because it can cost up to 50% of the value of a material to fund and administer its being stocked, year on year. Stock is an absolute evil; reducing it brings benefits that are rarely understood or properly costed.

9.6.2 The essentials are to make the product attractive

There's no point in making it cheap to produce if customers won't buy it because they don't like it. You will have wasted countless time and money that you painstakingly committed to previous stages of the process. It's amazing how many times this mistake has been made in the West; so many attractive concepts have been productionised into sales disasters.

If appearance is a key buying factor, do customers think that your product's appearance justifies its price? Don't impose unimaginative *adaptation for production* that sabotages what the industrial and engineering designers have taken great pains to achieve. Customers do not appreciate cheap and nasty products. They can buy that elsewhere at very low cost. Your business is to increase perceived value in up-market products so you don't get competition from low–labour cost countries. The production engineering detail must have been developed in partnership with the production engineering part of your team during the concept phases, not applied as an after-thought or because departments or personalities don't want to collaborate.

Graphics and literature

In your promotional support literature, you identify the key and unique features that trigger customers' buying motives. It's not about what you think is fascinating engineering detail. How will it solve their problems; what will it do for them; how will it benefit them? Graphics and literature must be integrated into the design process of the product itself because they form an essential part of what you are offering. The product offering includes all its related literature. Poor graphics and support literature, of all types, can ruin the customer's perception of an otherwise excellent product and of your company.

Design not just the product graphics but also

1. packaging and protection for transport;
2. the operating manuals and instructions;
3. manuals for service and training;
4. sales data and literature;
5. launch and point-of-sale material;
6. a customer canvassing response card;
7. the guarantee document and certificate.

9.6.3 Simplicity of detail is vital but difficult

Parts variety

> It takes a far greater intellectual effort to design a product that has fewer parts and is simple to manufacture.

There has to be a constant focus on the need to reduce parts' count. Aim at fewer components in each succeeding product; lower parts variety and easier, faster and more reliable assembly. This is vital for both cost and reliability.

It's easy to create a complex product that has many parts to meet a functional requirement. In some cases this has lead to using advanced manufacturing techniques that would have been unnecessary had more intellectual input been spent to make the product simpler. The effect on capital investment and manufacture cost can be dramatic.

The more parts your product contains, and the more complex they are, the more your potential failure rate will rise. You need parts and manufacturing processes of exponentially higher reliability and quality as parts' count increases (because of the interfaces between the parts). For example if you manufacture a product that contains 40 components, and one component in every thousand is faulty, 4% of your products will show faults. With 400 components, it rises to 35%. To produce motor vehicles where fewer than 0.1% of the vehicles exhibit a fault requires the components to be supplied with fault levels of one part per million or better. It is feasible to aim at zero defects; Shingo[10] explains how.

You also need to analyse your warranty experience of the carry-over parts you're using to make sure you iron the bugs out and don't incur the same warranty failures on your new product.

Interfaces

Inadequate standardisation of methods will increase costs and can decrease reliability. Ideally, you design the interfaces between building blocks so you can use new with existing ones.

[10] *Non-stock Production*, Shingo.

Interfaces between building blocks or systems are important because the more there are, the more likely you are to develop problems from unexpected interactions between them. All this should have been sorted out in optimising the concept, not left until now. In some cases, despite the benefits of re-using systems from other products, it can be less costly in the long term to develop new combined systems to eliminate interfaces. It will depend on the complexity of the systems and the problems you encounter; it needs logical analysis to arrive at the decision.

Simplicity lowers costs

That is why it is important to aim at fewer parts, both in variety and parts-count, and fewer system interfaces. Aim at simpler machine set-ups. Reduce production costs by combining functions into fewer parts; design them to be easier to manufacture and assemble. It is often better to design a single more costly complex-to-make part in place of several others. One UK machine tool manufacturer designed the functions of 36 previous parts into one and, although the resultant casting was very much more complex, the finished single part cost 50% less than the previous 36.

Simplicity and customer choice

But reducing complexity is not about reducing customer choice. You use the launch specification as the control; it will (or should) contain the detail of what the customer wants the product to do at concept and functional level (see previous phase).

> Reducing complexity is not about reducing customer choice.

The ground rules are simple: does the customer want it; will they pay for the features – your package? Don't reduce features that will affect your customers' perceptions, such as visible features and performance. Instead, work on the invisible or subtle features that they don't immediately see or appreciate. Go for minimum variety and maximum commonality on the invisible, not obvious systems.

Parts proliferation

Review how new components are designed and catalogued: advances in the automation of design techniques can often make it too easy to design a new part than to re-use one. This increases cost not just through production cost but also through spares holding and servicing. That is usually because it can take too long to search for and identify a suitable existing part; in some cases the task is so daunting that it's almost impracticable. Develop a counter-measure or you will end up with a parts quagmire that could exponentially increase your costs. This has been a major contributor to driving some old established companies to the wall.

Descriptive parts numbering systems are one way of avoiding this problem. This is especially important in firms that have a long history of developing families of products where historical parts variations can become an out-of-control nightmare. If you are not vigilant, almost identical parts will have been designed at different times by different people; that's where a descriptive system, discipline and understanding the objectives helps.

Developing unplanned variants – customer specials

New product development is an investment for the company's future. So all variants must either be profit making, chargeable to the client or maybe a loss leader if it leads to further opportunities.

A new product needs the full product development process. But creating an unplanned variant may not – there is clear distinction (see section 5.15, Project types). Separating these two activities can be important to minimise cost.

> A new product or a planned variant needs the full product development process. But creating a customer special may not – there is clear distinction. Separating these two activities is important.

To do this, some companies create a small support group. They interface with production, are close to production problems and to the release of new products into production – a gateway between new product development and production.

As TSS Ltd found: 'We used to have only one form for all new product developments. There was a backlog of about 40 projects. But only two were new products that needed the complete process. The rest were customer specials. So the new group now quotes for and project-manages all customer specials. This has had a major beneficial effect on our long-term programme.

In the past, every time we had an enquiry for a special, engineers were taken off important critical projects for a few days. They would consider the special, write a specification for Sales, and do all the planning to make it happen.

Now, the support group engineers project-manage them all. If they determine that the special needs say three hours of software time, they can arrange to use the software engineer for three hours. This avoids major disruptions to the new product programmes. The engineers working on key projects no longer have to deal with a constant stream of minor interruptions.'

9.6.4 Safety aspects

Safety means more than human safety. It includes situations where the failure of one component may adversely affect the performance of another and lead to operational problems whether or not human safety is involved. While clarity and simplicity automatically contribute to safety in the product, there is an established safety hierarchy, which in priority order is

- direct approach: eliminate the hazard;
- indirect approach: apply protective systems;
- provide warnings;

- provide for training and instruction;
- prescribe personal protection.

9.6.5 The significance of suppliers' contribution

Your product's merits will be limited by the product development expertise of your suppliers. In some cases, your suppliers may have designed more than two-thirds of your product: you are therefore compelled to make use of their expertise.

Are you using the strategic advantage of your supply base? What is your long-term component sourcing strategy? Your component suppliers should have contributed the information they possess on the performance of their parts and systems at the concept stage. If it was deficient, are there better suppliers that understand their products? Or, if they are strategically irreplaceable, can you run programmes with them to help improve their capability to match your needs? It's rather late in the day to do much about it now: this is planned as described earlier in the concept stage.

Bills of material

These can be used to initiate purchasing development, with supplier input on early production bought-out parts and systems.

9.6.6 Manufacturing instructions

Prepare detail drawings and design documents (or the equivalent for your company) as manufacturing instructions:

- drawings for control purposes in production;
- drawings for supply scheduling;
- parts geometry interference/clashes/incompatibility;
- limits/tolerances in collaboration with process capability studies and with service simulation testing;
- parts explosions, bills of material, build lists.

Your studies on robustness (in Phase 4) will have identified the critical areas that have to be specified precisely to achieve design performance.

For the rest, beware of your staff sprinkling tolerances and specifications like confetti all over your drawings. It adds extreme levels of cost. Many tolerances may be unnecessarily tight and some may be impossible to achieve by the processes you operate. For example you might have to use grinding where turning would have sufficed, or an electronics component of far higher grade than necessary for the performance. Some parts may become almost impossible to manufacture or obtain at all.

To avoid this, your production engineers need full knowledge of your own and your suppliers' process capabilities.

In detailing every part, specify materials, finishes, bulk and surface treatments. Assess the availability of the materials specified. Most should just be confirmation from the concept stage.

Cost targets

The project leader needs to act as cost baron to adjudicate trade-offs between areas where they could lead to major total reductions.

The aim is to minimise overall cost and this does require trading between areas. Set a cost target for each sub-system:

- allocate responsibility for achieving it to the group responsible for designing it;
- monitor their progress to achieving it as part of the project's control programme;
- continually seek ways to trade costs between areas to reduce the total.

9.6.7 Final production engineering; plant acquisition and commissioning

Define the critical process-control points in the production sequence to ensure the required performance, as identified for robustness. Your production engineers would best be working to principles such as those set out by Shingo[11] to achieve the lowest costs and highest levels of quality.

Production engineering is done concurrent with, and as part of, the concept and detailing stages. Much of the comment is covered in those sections. The production engineering work will have proceeded in parallel from the earliest concept phases. You will already have schemed an outline of how the product will be manufactured. You progressively develop and confirm the final detail as the design develops:

- the facilities and the methods you use for the product;
- its design for manufacture and assembly;
- process and materials planning;
- resource planning;
- assistance in reducing costs;
- process and cell balancing, tooling system design;
- setting out the manufacturing standard operations;
- make or buy: the contracts finalised with suppliers;
- acquisition installation and commissioning of plant.

Much of this will have been able to have been ordered in earlier phases where final detail is not necessary. For example a machine tool is rarely dedicated to only one component (see comments on 'monuments' in *Lean Thinking*).

[11] *Non-stock production*, Shingo

Materials selection; engineering components and raw materials; surface engineering

A host of websites gives useful information on these and many other design aspects.

Product development

www.mynott.com/tics has continually updated information on useful publications, case studies and events on product development, with links to relevant websites.

9.7 Phase 7: Early production and confirmation trials

Produce, test and evaluate early production units to confirm their performance characteristics and to verify the production processes. Ramp up production capability. After a time, do a post-project review to capture lessons.

9.7.1 Confirming product performance

This is the first time you have a product that is manufactured from production tooling and processes. These are not development trials but a confirmation of what you already know, both of the product and its means of production. So you don't manufacture these early production products from a special facility by methods that don't represent the properties of your ultimate production. There's no point in doing that because trial products with unrepresentative components won't represent reality: there's no point in making them.

This is not where you build pre-production prototypes to find out whether your product performs or not.

You should already know that from your feasibility and performance testing of concept building blocks in Phase 3 and of optimised prototypes in Phase 4. The aim of this stage is to confirm that your product will perform as you predicted, by confirmation tests on early production products that you subject to service conditions and approval tests.

This is your early production stage. You build everything off production tools and processes – on the production facility you have installed for the product. You use the exercise to do final confirmation trials of the production methods and equipment as well as of the product. It will confirm their performance and reliability, as well as those of the product. If you've done everything in the right order, you certainly won't require tooling changes. If you do, insufficient work and testing will have been done in earlier phases.

Build a small trial production run. When you start doing things this way you may gain confidence if you build two or three progressively larger subsequent runs, to finalise production and product final fine-tuning. When you are thoroughly familiar with this methodology, only one trial batch should be necessary.

Legislative and approval tests

Use these early production product trials to confirm the performance of the product against market and in-house criteria and duty cycles, to confirm what you should already know from your having devised the final concept design scheme. Use some of the products for the confirmation tests; some for legislative and approval tests; some as demonstrators; and ultimately sell most of them as your first run of the new product. Much legislative testing may have been possible at the close of Phase 4, using optimised prototypes built with as near as possible off-tools properties. In this phase you complete what was not possible then.

Final changes

Modify the design details only when absolutely essential – little should be necessary. Some changes might be able to be left to your next update. You should not need to make many, if any, changes before you start full production. Following this phase, you will be continually improving the product and its production methods by systematic small improvements throughout its life in full production.

9.7.2 The transition between product development and production

Your product and production processes will be ready. Your product development team will be making final preparations to monitor the product's performance in customers' hands. They will have planned their response to likely customer problems. They may also be preparing the business plan for Mark 2 of the product.

Final preparation activities include

> Use a formal production parts management system that makes sure beyond any doubt, by methodical checking, that you are completely ready to start full production.

You thereby avoid discoveries that certain parts have not been sourced, or that parts have been ordered to wrong drawing issues – the possibilities are endless and can cost you lost sales. A formal system repays the investment in time and money many times over.

Other activities will include the following:

- continue to monitor technical user and market trials;
- ensure that all the production-issue detail drawings are issued;
- that suppliers are supplying to the correct drawing versions;
- packaging design, manuals and all literature are available;
- the route to market is ready, the distribution network in place;
- promotion is ready to run;
- final de-bugging of production equipment and tooling is complete;
- all process specifications, training and documentation are in place.

Your marketing effort will be under way:

- start production, distribution and sales;
- monitor feedback from sales;

... and so to further products ...

- prepare to anticipate your market peak by planning to launch successors or stable-mates to improve your market share.

9.7.3 Post-launch activities

Product tracking and maintenance require

- review of sales, service problems, customer reaction;
- tracking, after-sales service and trouble-shooting;
- monitor warranty problems and put counter-measures in place;
- monitor and improve costs and manufacturing performance;
- review sales, dealer performance, appropriate for your firm;
- continuously monitor customer satisfaction;
- research and monitor competitor reaction.

Information to customers, long-term service and technical support, spares, re-cycling of components and materials. Complying with re-cycling requirements; can the product be dismantled easily? What are the lessons to pass on to Mark 2? Update the history drawings (see section 9.3.2).

How you measure results – 'metrics'

How do you measure your company's product development performance? What company targets will you set?

Inputs, such as the percentage of revenues spent on product development, numbers working on programmes and so on, are important. You need to know these to ensure long-term stability of resourcing. Large year-to-year fluctuations, or gradual attrition, will lose the knowledge base you have built up in your staff. It bins your investment in the key competencies. But input measures won't reveal the effectiveness of your spend.

You monitor output measures to see how effective your product development programmes are. And not just financial; you can have brilliant financial performance but might be losing market share. You monitor

- market share trends of successive new product introductions;
- proportion of revenues from products less than 3 years old;
- the revenue from products introduced in the past three years compared to their product development costs;
- the value of expected future revenues from projects under way compared to their expected product development costs;
- your suppliers' product development performance scores;

- your time to money (rather than to market);
- quantity of useful suggestions during development.

You select measures relevant to your own product sector. For example in the pharmaceutical sector, it could be important to monitor the number of patent applications; in the consumer electronics sector, the proportion of revenues from products introduced during the past 12 months – and so on. Do not just use financial measures – you will be misled.

Your budget: the division of product development resource

You will not be spending 100% of your product development budget just on developing new products. You will divide spend between a number of areas. A typical split might be

Background product development	5%
New products	50%
New applications of existing products	15%
Enhancements to existing products	15%
Continuous cost reduction exercises	10%
Engineering services to other areas	5%

The reason to set this out is to ensure that you resource all areas. Usually what happens is that the new developments get halted temporarily while other areas are serviced. The result is a continually interrupted mainstream programme, with launch dates missed and over-run budgets. It is better to realise that these relatively minor 'other projects' need resourcing. Produce a realistic plan, rather than continually interrupt key projects – which is what happens in too many companies. See sections 9.4.4 and 9.6.3 on variants.

Design 'maintenance' can absorb 50% of your design resource

How will you deal with the design changes to remedy warranty concerns? This can absorb a lot of resource. Some companies temporarily recall members of the disbanded project team to do this. But you must investigate why so much effort is necessary: is it because tasks have been omitted or poorly done? How will you reduce this in future?

> Why do you need to spend so much resource on post-launch design changes?

Chapter 10
Running and managing the programme
How to set up and manage a project

How you set up and control each programme will determine its success. What targets should you set and monitor? What is financially important and what is not? How do you detect potential time over-runs and take countermeasures to avoid them? What formal processes do you need or not need? How do you manage multiple projects? And how do you assess risk at the outset and as you progress?

10.1 Introduction

Justifying expenditure usually involves quantifying returns. This is a minefield, subject to more creativity than found in the Louvre. The hypothecated returns will usually be increased sales through earlier introduction, better margins, reducing the work needed or improving the product leading to better sales and lower warranty costs. It all depends on how well you manage the project.

> Successful projects depend on two key factors: having the right culture and having the right process. One does not work without the other.

How you organise and set up your project culture is discussed in *Chapter 5*. How you set up the right process is discussed in *Chapter 9*. This chapter deals with how you control the total project. You have to ensure that the outcome generates profit. There is no other reason for the project. But ensuring it makes money is not the same thing as tightly controlling expenditure. Understand what creates the money. Generate a financial model to ensure you control the aspects that will earn money, not impose controls without understanding their effect (like most companies do).

10.2 Project management

Project management is about people and the way they are organised, managed and motivated. It's how you achieve your profit targets. It succeeds only if this is recognised, with the climate and structure where everyone takes responsibility. Only when you get this right can you use the tools and techniques effectively. But most companies do it badly.

Project management is essentially about people and the way they are organised and motivated.

Changing any system is a socio-technical task. Unfortunately, most of us spend 99% on the technical part and almost nothing on all the 'socio' activities that will make it work, like developing collaborative initial requirements, feedback, training and so on. As an example, usually only 20% of all new software projects meet all their requirement; 40% fail and the rest don't meet all the objectives. More examples:

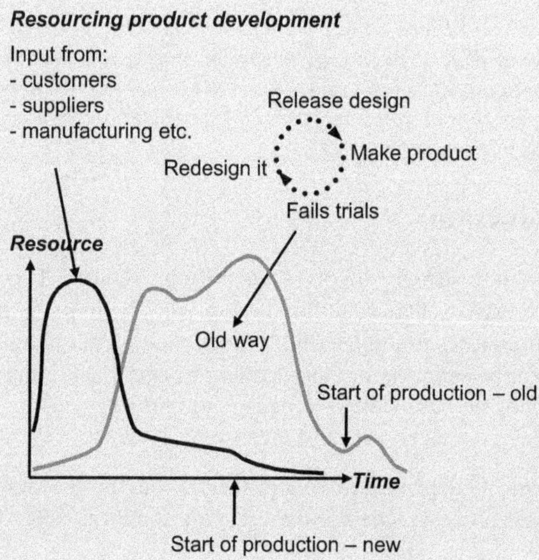

In 1996, the EU ESPITI project reported on 8000 European companies' software development projects: only 20% of them completed on time. The majority were at least six months late and a quarter were cancelled.

Engineering magazine's 1993 survey of more than 500 UK engineering companies showed that more than half did not complete their product development projects on time. The average time over-run was over 25%. Not only that, but the average company took 20% more unbudgeted development time after production release to get the product right for manufacture, and incurred 25% of their engineering changes during this period.

This is all symptomatic of inadequate definition at the outset of the process. The figure above, amplified in 9.2.6, demonstrates why. Inadequate early work is costly, kills profit and loses market share.

In addition, the best engineering practices are not widely used. The key role of product development staff, especially design staff, is often not well understood by top management. Many take on a wide range of new activities to support product development but management often fails to recognise the significant effect of these extra activities on the final result. Traditional cost accounting does not evaluate their contribution to profit and therefore cannot compare good practice with poor.

10.3 The operating model

If you don't have a model process to follow, you can't plan the tasks and control progress. Then, if you don't use the appropriate tools and techniques you won't get the best outcome.

Chapter 5 considered ways to organise your company. But your internal culture is more important than how you are organised. How well people collaborate with each other at all levels; how open minded everyone is, how blame-free the culture is to learn from mistakes, and how willing everyone is to follow the process and not buck the system. These are the minimum requirements that make it easy to control your projects.

It can take from 4 to 8 months for a company to design a good operating model and up to four times to sort out the culture to make it operate perfectly. The main retarders are management's beliefs and values – principally their decision-making behaviour. And laziness in not doing the work diligently enough, especially in *Phases 1 and 2*.

The easiest way to put in a new system is to enhance the old one. And once in place, it needs regular review to respond to the lessons learned, and to avoid complacency. Continual training helps avoid this.

10.4 The programme – priority and targets

10.4.1 Your company's financial model

> Having a brilliant process does not guarantee a brilliant performance. The other half of the equation is company culture. One does not work without the other: you need both.

You would expect to set programme targets to be able to launch the product as soon as possible at minimum cost. But the four key requirements for a product development project are

1. cost of the project (usually directly related to man-hour input);
2. time schedule of the project;
3. unit cost of the product;
4. performance of the product.

You need to model how these four requirements influence your profit. It will then become clear which you should concentrate on controlling, and which are less important or quite unimportant.

Finance managers are trained to control project cost irrespective of how it affects the other three factors. But tight control of project cost does not maximise the final profit. The other three do, but it's a trade-off. Which is most critical will depend on the type of market you are aiming at. In highly competitive newer markets, time to market is the key. In more mature markets where product lives are far longer, unit cost is likely to be.

Smith and Reinertsen[1] show how you produce a model of the process and calculate the relative significance of these critical factors. The main constraints are

- time schedule;
- scope of work;
- resources.

In real life you trade-off between the three depending on your objective. The usual solution is to control two closely and give the third leeway.

10.4.2 Balancing priorities

> If you don't have a financial model of your programme, you will not control the right parameters to minimise time and cost.

Prioritise the project, review and control it to what is most important:

- delivery date; increase resourcing if required;
- project cost; this has priority over timing;
- developing the technology; solving the problems has priority over time-scales and cost; however, you should not put intractable problems into your main project time lines – resolve them in your background development programme.

You can use this outline to judge the ramifications of new challenges. For example, should you let your project's timing slip to achieve better functionality and quality? In each case, you have to examine the effect against your financial model and your target marketing specification. If you want to add features, why aren't they already in the marketing specification? Is just someone's suggestion or a fundamental new 'discovery'? Will omitting the features reduce sales? Why has it only just been discovered? Will the added features generate sufficient extra margin or added sales to compensate for the revenue lost through extra development costs and late introduction? Or should you hold the new features for Mark-2, later? Let your financial analysis guide your decision.

[1] *Developing Products in Half the Time, Chapter 2*, Smith and Reinertsen.

Your plan

To monitor progress continuously requires a formal sequence of product development tasks to be set out beforehand – a plan and a set of rules, with project milestones at decision-gates between major phases. Tasks are overlapped (see section 10.5.2).

Subdivide individual tasks. The longer they are, the more they are likely to slip before you notice. And the more difficult it will be to bring the programme back on time. Some companies aim at a 2-week maximum subdivision in a 12-month project. Longer than this, and the slippage becomes less visible and too costly to correct.

Special challenges from the concept phases

How long will it take? To schedule your programme you need information on the man-hour needs of each phase. Concept phases pose the most difficult estimating challenge. Don't just set an arbitrary time and expect the programme to meet it. Even crude estimates are better than arbitrary figures. Just guessing may lead to wildly missed timing; or it may create queues for facilities from out of synch programmes. Or worst of all, it may lead to a degraded product because you have to trim what is possible to meet too short a remaining programme time.

You can base your man-hour estimate on measurable parameters such as the number of expected components or by seeing how long products of related complexity took in the past.

Devise the detail programme

First devise the master programme: set out the major milestone stages with target times and costs. Then subdivide the milestones into manageable major tasks. Assign each major task to the team that will have responsibility to complete it; get them to work out their own detail tasks, resourcing, schedule and timing. Within the master programme, get them to set their own 'local' programme; they own it; they run it. You tutor them, monitor it with them and help remove obstacles.

Make it a continuous culture where the teams devise, own and run their own part of the programme. But don't abdicate responsibility. The purpose of their activities is to achieve company objectives, not carte blanche for team members to do whatever they like. It has to be monitored.

Continually reduce time and cost targets compared to previous projects.

> You drive the project, from idea to product launch, by defined phases initiated by formal reviews. You check that all deliverables are available to start the next phase. And you define the transfer criteria for the next phase.

Represent the process as a set of simple activity charts on an A2 sheet and post it on the wall in each area. A slim manual can answer detail questions on any individual activity. And beside the total activity chart, every week post an updated GANNT chart for that area's programme. The two together will make the process and progress clear and visible to everyone. Hiding them in computers and weighty manuals (footrests and doorstops) is a waste of the effort you've paid for.

10.5 Controlling the project by reviews

Drive the project, from idea to product launch, by defined phases, each preceded by a phase review, or decision gate. Many different terms are in use – stage-gates, decision-gates, review.

Define tasks and sub-stages to break the project into short phases that form a sequenced process. Each phase needs several functional departments of the company to work together to complete.

The gates are reviews between phases to control the project. See 'Progressively reducing the risk' chart and text after the Preface. Each review uses a checklist of items (transfer criteria) to confirm the 'go' or 'no-go' decision. The purpose of the review is to check that all deliverables are available to start the next phase. You also define the transfer criteria for the next phase, and check they are clear and understood by the project leader and team.

10.5.1 Phase reviews

Check for interaction between projects so you don't overload the system. Try not to review different programmes at the same time or require the same people to do priority tasks and double-book their time.

> Do not run the project as a series of complete phases with a halt at the end of each while progress is reviewed; some parts may want to make faster progress than others.
> Waiting for decisions is a major factor in schedules slipping: therefore, move to a process of continual decisions.

Phase reviews may sometimes find that the required deliverables will not be able to be produced, either on time or at all, and the project will need to be re-focused. The schedule and resourcing then have to be revised to bring the project back on time, usually by devoting extra resource or exceptionally by revising the specification – but only if customer-justified.

In extreme cases, the project may have to be cancelled, where problems of achieving time-scales prove insurmountable. A cancelled project is not a waste: the information will be incorporated into its successor. This results in better understanding and better-organised future projects.

10.5.2 Concurrent operation

The principle of controlling projects is to review the completion of one phase before starting the next. But that doesn't mean that you wait for every part of the phase to complete before you approve starting the next. Doing it that way will cost you time delay and lost sales revenue. This would be akin to manufacturing in batches – slow and costly.[2]

Some parts of the project may need to run at different speeds. So subdivide the project into sub-projects, each a functional sub-system with its own definable boundary, which can be completed and approved separately. For example, you develop a car's engine, drive train, body structure, interior trim, exterior trim and so on as separate parallel sub-projects, with sub-sub-projects within each. You don't wait for every sub-project to finish its phase before starting the next.

Let each sub-project run at its own speed with a series of small, fast, sub-phase reviews to avoid delaying work that could progress faster. And when all the separate sub-phase parts are complete, the total phase is signed off.

> The control of interfaces between systems becomes vital if each runs to a different time schedule.

The two major constraints are

- You can't start doing this until the marketing spec is complete (end of Phase 2); and they all have to come together at the end of Phase 6.
- And second, control the interfaces between sub-project systems.

Plan to use information as early as possible, for example, where a part of the project has a long tooling lead-time. You don't need every scrap of information before you take action. If it's clear, having schemed the concept that a component common to all concepts will require large press tools, right-size blocks of tooling material can be ordered before you select the concept and define the final shape. You just need to know the size envelope. Then, as the shape is defined, material can be roughed out of the blocks, progressing until you know the final shape.

And if new equipment is needed, a press can be ordered as soon as you know the tooling envelope size, panel complexity and likely tonnage.

And you don't wait until the conclusion of *Phase 5* to sanction capital expenditure. You approve an outline total CAPEX at the conclusion of *Phase 2*, and call it off as required during *Phases 3–6.*

This methodology has become a major factor in speeding the development of products – it can more than halve your total lead-time. It has been practised in part by carmakers since the 1930s. But the difference now is that product teams with members from all functions are used. That makes for fewer changes and further reduces time because information is provided early that previously would wait until the project reached the department concerned.

[2] *Non-stock Production*, Shingo.

10.6 How many reviews and phases?

As Brian Cooke, formerly Director of Engineering Black and Decker Consumer Power Tools, commented on programme management:

We introduced a milestone process in 1994–5. Each project's programme is unique. The programme manager, appointed before the start of the project, puts it together. Each one is tailored to the proposed product's degree of newness, the areas of technical risk, and so on. This determines the number of management reviews.

Everyone involved in delivering the technical programme knows the dependencies between the commercial and technical tasks. This has now been made visible. All the engineers understand the importance of working prototypes to the marketing group. They understand the importance of the marketing group having pictures of products so they can create brochures, point of sale literature, and so on. They understand the importance of getting the right competitive test information to the marketing group at the right time so that literature and advertising briefs can be compiled.

The linkages between the technical and commercial programmes are now well established whereas previously they weren't. We now understand how long it takes to develop programmes and why it takes that long. We understand which elements are on the critical path and which elements we need to shorten to reduce total time. We make continuous evolutionary progress.

The programme managers are key to this process. They are normally drawn from the engineering community and are commercially minded engineers. We make sure that all our senior engineering managers are exposed to the commercial realities of running the business.

The number of reviews you need will become more evident from your initial risk analysis (see *Chapter 10.6*). Higher levels of risk will be associated with newness of a number of aspects, be they technical, commercial or financial. The higher the level, the more detailed your control strategy. For low-risk projects, many of the phases can be merged. For example, a modification to an existing product will need far fewer reviews than a totally new one. As in most other aspects of product development, you tailor the control to the nature of the product and its markets.

But if you decide that the risk level justifies fewer reviews, beware of the pitfall described at the end of section 10a.You must answer all the checklist questions of the reviews you have merged, not just those of the actual review.

10.7 Aspects of financial control

Approve projects according to their risk–reward profile, giving priority to the most viable. Where a number of projects compete for resources, the system put the weak projects on hold, that is, those that will contribute least. These can be archived for resurrection later where conditions develop that enable them to become competitive.

Assess what blend you want of low-risk (perhaps incremental) projects and higher risk totally new projects. (See section 10c on assessing risk.) You can set different return objectives for each (see section 10.7.1 and 10.7.2). You can also use the system to justify financially the acquisition or divestment of product lines.

10.7.1 Using IRR to assess project viability

Compute each project's internal rate of return (IRR) periodically. It can be calculated more accurately as the project progresses and as the unknowns are resolved. Soon after their start, you will begin to see a trend for each project. Where it becomes clear that this trend will never achieve an acceptable return you can revise the project or shelve it.

In the later stages, you can use actual costs and contributions to monitor the product's financial viability in real time and compare it with other products in the company's range. This will make it clear where resources should be applied to give the best payback.

> Periodically compute each project's internal rate of return.

You apply different internal rates of return targets depending on the nature of the project. These can be variable financial hurdles to reflect the level of innovation and therefore the risk. For example one company in an expanding market used

- 10–20% for incremental change to an existing product or for replacement of essential tooling.
- 15–35% for a new product similar to an existing successful one (little innovation and well-controlled risk).
- 40–70% for a completely new product (high innovation and most risk).

10.7.2 Computing the IRR

The item lists below are just examples that you supplement with your own inputs. A number of model programmes are available, for example those published by Hewlett-Packard in some journals.

To calculate a project-based internal rate of return, you need to identify at least the following costs and contributions that can be attributed directly to the project. Initially use estimates; update them with actual costs as the project proceeds. The items are readily visible from the tasks you have specified for your programme. Examples are

Phases 1 and 2

- licences for ideas;
- market research;
- pre-development and marketing test work;
- detailed investigations to prepare a business plan.

Phases 3 and 4

- background development of concept ideas;
- consultancy inputs;
- modelling and trials to prove product concepts;

- simulated/virtual test work (product, market or technology related);
- manufacturing studies;
- final concept selection;
- develop prototypes to demonstrate and test (internally and externally) every aspect of the product.

Phases 5 and 6 – detailed design and development

- preparation of launch specification;
- prepare a project CAPEX with IRR and sensitivity analysis;
- detail design
    - *production release, tool up for production*
- manufacturing costs;
- factory layout, assembly jigs, machines, test rigs, tooling costs;
- literature, packaging and merchandising costs
    - *production start*
- material and labour costs;
- market tests using early production examples;
- approvals testing and certification;
- sales training, advertising and launch costs.

Phase 7 – product life costs

- project team support for ongoing production within specification;
- sales for years 1, 2, 3, 4, 5, 6 etc.;
- sales promotion costs;
- patent and copyright costs to protect IPR in relevant markets
    - *maintain product specification*
- investigate and justify change proposals;
- implement changes
    - *eliminate the product*
- investigate an elimination proposal;
- total project-related obsolescence costs.

10.8 Conclusion: continuous development

At the conclusion of any project, review the lessons learned and update your manual so that subsequent projects incorporate the improvements. In particular, identify the features that really helped to create success so that the next project adopts them. Essentially you need to analyse

- What was supposed to happen?
- What actually happened?
- Why were there differences?
- What can we learn and do differently next time?

Part a Programme control reviews

Monitoring and controlling a project requires a financial and an operating model. You control the project phase by phase by formal review and you run the reviews in a way that minimises time-wasting argument. You run different systems' phases at different speeds to cut time.

10a.1 The review meeting

The purpose of the review meeting at a milestone date is to review progress and to authorise whether or not the next phase of the project is permitted to proceed. (See also section 10.5.2.) Progressing to the next phase is not an automatic rubber-stamp decision; it must be authorised only if the deliverables have been produced or can be guaranteed to be produced.

There are three possible outcomes of a review

1. approval to continue;
2. approval conditional on satisfying certain conditions;
3. approval withheld: the programme does not appear to be viable – terminate it.

If a project looks as though it will not achieve an acceptable return on assets – or whatever criteria the board sets – it should be terminated and its lessons analysed so they can be used on future projects.

10a.1.1 Scrutinise at each review

At each review, the project is scrutinised by the review group who seek answers to a number of general questions and who examine the answers to a set of **transfer criteria** specific to the phase or sub-phase just completed. Transfer criteria are items that need to be verified before approving continuation. If the project is given approval to continue, the review group sets the transfer criteria for the next review. They may also allocate priorities. Typical transfer criteria are set out below as a starting point to generate your own: use brainstorming to create lists for each sub-phase then refine them.

> You monitor a project from start to finish to make sure it remains viable and will satisfy its objectives. This is done by pre-determined reviews.

You choose your milestone points to hold reviews to suit the way you run your own organisation. The programme for a completely new product could have reviews at the start of each of the seven phases. But for projects that are alterations to existing ones or minor modifications, you would combine phases and their reviews.

The review board must have the power to make decisions on the spot, not refer them to another higher-level meeting at some future date. That is a recipe for increased cost through generating delay.

Nothing can cause more delay than misunderstandings; a good project manager will ensure that everyone understands what is required of them and by when. This covers all aspects including help to interpret design changes.

Those presenting to the review board have the responsibility to report on the outcome of each activity under review. Each person responsible for a particular task will have collated their results before the meeting. If the task is not 100% complete, they will be responsible for planning and presenting the necessary counter-measures. They must ensure that the counter-measures are in place before the meeting.

To the question 'Is the task complete?' only one of two answers is permitted:

1. 'Yes, the task is complete in all detail'; or
2. 'No, it is not complete; here is the plan to complete it in full detail by (date); we have consulted (areas x, y etc.) involved in the plan; they agree their part is feasible and we have their commitment; the activity will be complete by (date).'

To achieve this, each area and sub-area responsible for presenting its progress report holds separate working sessions in advance of the meeting. To ensure that the programme will not be delayed, each group must devise and test a counter-measure to any tasks that are not completed. If proposed solutions need collaboration with, or have an impact on, other areas, the proposed counter-measures must be checked with these areas and any problems sorted out. At the meeting, you don't want an area that is affected but has not been consulted raising objections to a proposed counter-measure.

The phase review meeting should therefore be short, well prepared and non-adversarial. Half an hour should suffice.

Phase reviews should be a fast, ceremonial confirmation that all is done, not an opportunity to discuss actions to remedy failure.

It is not a debate, an exhibition of innovative excuses, a demonstration of finely honed inter-departmental battle techniques or a political debating chamber. Nor are phase reviews working meetings where 'discoveries' are made. Or meetings where you hold discussions to solve problems; those present often don't know the technical detail well enough to propose solutions. This needs problem-solving techniques and methodology – by a small expert group prior to the meeting. Making discoveries or attempting to solve problems in these meetings shows poor preparation and poor project management; it wastes company time.

A 'no-blame' culture is essential to prevent time-wasting politicking and buck-passing. Full disclosure of errors must be encouraged, not punished. On the other hand, if you find that presenters deliberately misled the meeting (they said they had completed tasks when in fact they hadn't), then take disciplinary action. If you don't, reviews become a sham.

In the extreme case, where you have not been able to pull back lost time, where the review is incomplete or unsatisfactory, you may have to consider moving the launch date.

At all costs avoid compressing the late stages of the programme. That could lose more profit through warranty costs, lost sales, and reputation, than losing revenue through launching late.

However, you can often launch on time but delay the first deliveries – that may do less harm. Either way, you have a problem; use your economic model to assess the financial effects and risks of the alternatives.

(Birchen[3] gives useful detail on how you standardise reports on meetings.)

10a.2 Review questions and transfer criteria

These are general questions the review group can ask to help them form an opinion on whether to allow the project to continue. The ones given here are common to most stages of a project. You create additional lists for each sub-phase. Typical general review questions might be

- Is the project still a good strategic fit with the company's future plans?
- Does the market opportunity still exist?
- Are all customer requirements met?
- Are the risks being managed?
- Have all alternatives been adequately explored?
- Are the resources available and in place for the next stage?
- Have all procedural, regulatory and contractual hurdles been cleared?
- Has the scope changed?
- Is the work sound?
- Are all transfer criteria therefore met?
- So do we still have a business case to continue?

10a.2.1 Transfer criteria

These are specific to the phase of the project. An example outline list of questions that the project teams need to address before the next review is given below.

Some transfer criteria may be answered more by opinion than fact, when the review group must assess the quality of the data that is submitted. After the

[3] *The Lean Toolbox*, Bicheno.

sub-phase has been approved at a review, the review group sets the transfer criteria for the next review. This sets the direction for the next stage, that will be peculiar to the programme tasks in hand.

An example list is as follows.

Example transfer criteria to authorise Phase 2:

- Does a market opportunity exist?
- Does the product programme fit our business strategy?
- What are the long-term implications of this programme to the company?
- What are the product's key features and does the technology exist or can it be acquired?
- What is the probability of technical success?
- Is the product over-ambitious; could a simpler version capture a significant market sooner?
- Are there any regulatory problems that could prove to be insurmountable?
- Are we obliged to continue in order to comply with a new regulation?
- Quantify the benefits: is it likely to generate an acceptable return?
- Are the benefits real – for example, will it generate additional business or cost reductions?
- Do we have the funds, resources and expertise? Would we need external help?
- Is there a simple way of doing it immediately?
- Has a risk assessment been made; are the risks acceptable?
- Is it a low-risk short-payback programme that can be fast-tracked?
- How many phases and stages are needed; define when should there be reviews.
- Is the programme's written definition and scope acceptable?
- Has the idea been evaluated previously?
- Has the review group been nominated?
- Has the project team been nominated?
- Is the work programme for all phases defined in outline
- Are all lessons from previous projects available for use in the next phase?
- Have the review body and transfer criteria for the next phase been set up?

If at the outset you decided to combine some reviews because the risk level of the project justifies doing so, beware of a potential pitfall. At the combined reviews, include the transfer criteria checklists from the skipped reviews in your combined review. Failure to do this may omit vital questions and lead to unpleasant 'discoveries' later in the project. As one chairman commented to me: 'We seem to do complex projects so well. So why do we always screw up the simple ones?' Answer: because they omitted vital questions at their combined review points.

If you would like a template set of transfer criteria for every phase, go to www.mynott.com, and order the manual[4] that goes with this book.

[4] Colin Mynott, *The Product Development Process – An Outline Blueprint*, ISBN 0-9538779-0-7 (available only from wp@mynott.com, £150, $300).

Part b Managing multiple projects

You develop the way you organise your teams to cross-fertilise strengths between projects.

10b.1 Why multi-project management is important

Most companies need to use multi-project strategies and internal technology transfer. This is almost a complete subject on its own. Cosumno and Nobeoka[5] describe this in detail, based on Japanese car industry analysis. Their conclusions are:

The key is to develop your resources and organisational capability. You overlay these skills onto the methodology set out in the previous sections. You need to link technical knowledge and organisation. First, plan your product portfolio. This is a key part of your company strategy.

It may take 5–8 years to develop the unique management capabilities needed; these are far more difficult to nurture than single project management. Few firms can do it and those that do achieve a unique competitive advantage. It's more efficient and flexible than just developing one at a time. It leads to better sales, lower costs and faster programmes.

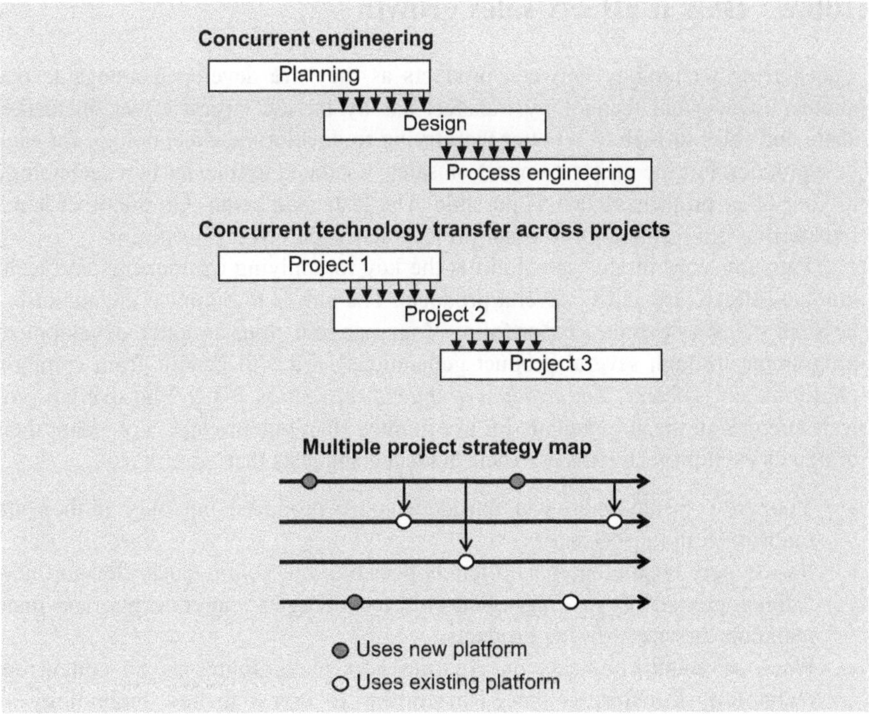

[5] *Thinking Beyond Lean*, Cusumano and Nobeoka.

10b.1.1 Lower costs and speed are the drivers

There is pressure to produce more sophisticated products at lower cost.

You will not survive just by efficiently managing one project at a time. But even with only a single product, you will need to transfer technology to your next ones. This in itself is a form of multi-project management.

In the long term you can't remain competitive even with a single niche product aimed at a restricted area of a single market. The market changes: user needs, market structure and competition. To survive, you need to develop a co-ordinated stream of products, and penetrate more sectors of the market. The key is to make the best use of your core components and systems and your technical knowledge. And continuously improve your product development management capability across all your projects.

Concurrent technology transfer *(transfer of technology between projects running at the same time) gives lower project cost than sequential technology transfer (transfer between projects running one after the other) because knowledge is transferred more effectively by face-to-face contact between the teams. Sequential technology transfer is better done by archiving data for the later project to use.*

10b.2 How it affects sales growth

Transferring technology between products as they are developed alongside one another (concurrent technology transfer) has by far the largest effect on market share and sales growth. It's better than trying to develop new technology for each new product. For the greatest impact on sales, you need to transfer new technology to your other projects as fast as possible. The faster the better. On the other hand, transferring old technology to other projects has little effect on sales.

Planning your product portfolio is the key to applying components and technologies effectively across several products. The aim is to optimise the attractiveness of the whole product portfolio. You map out your product development programme through several product generations so it will benefit from common platforms and systems. See *product architecture*, section 8.3.2. You overlap projects and design the components for use in more than one product. You adapt their design as each project evolves. Their evidence suggests that

- You grow faster when you introduce more products; but only if they are innovative and up-to-date.
- Re-use core technologies as often as possible; but continuously develop new technologies: don't just regurgitate old technologies and concepts; you must maintain 'newness' in the products.
- Place as much emphasis on creating new technologies as on concurrent technology transfer; you need a strategy to invest in new technology as well as to retain and transfer existing; you may need an independent technology team to do this (see *Background product development programmes* in section 9.3.7).

- Concurrent technology transfer between products reduces engineering hours by reducing rework on components and systems for the new models. But you must plan which components to share; co-ordinate doing so across projects and functions.
- It is far more cost effective to plan to re-use product platforms before the base project starts than to decide to do it later.
- Developing only totally new designs need longer lead times and more resource. Introduction is slower and 'newness' in confined only to the new product. As a result, the range does not gain as much extra sales.
- Customers do not respond well to old technology even with complete styling redesign. Continually modifying existing designs without injecting new technology loses 'newness' and customer appeal.

Cusumano and Nobeoka found the following results from typical strategies:

	Effect of strategies			
Strategy	**Lead time**	**Engineering hours**	**Market share**	**New product introduction rate**
Concurrent technology transfer	1.0	1.0	1.25	1.00
Sequential technology transfer	1.0	1.5	1.10	0.74
Totally new design	1.2	1.9	1.05	0.65
Incremental design modification	1.0	1.7	0.85	0.68

10b.3 Multiple project organisation – department or matrix?

Create separate projects that follow on from the base project. Ensure that the teams feed technology one to the other as the projects run side by side. This avoids loss of product uniqueness through poor design compromises and you won't waste time on unnecessary co-ordination.

> Avoid expanding a single project to accommodate variants: you won't be able to manage it.

It's difficult to mix project and functional departments; you need co-operative people and good organisation processes. It's even more difficult to manage multiple projects. In a matrix organisation, who has the final say? Project manager or department head? Company culture is the key.

What are your aims:

- to optimise the technology for a particular product building block (or component or system)?
- to optimise integrating components or systems in the product and achieve lowest time and cost?

A number of factors affect the outcomes, such as the need for

- fast rate of technology change (needs departmental expertise);
- shortest possible project duration (needs effective project teams);
- best interdependency of components (needs organisational expertise).

Project-based organisations are better at global problem solving; at innovating by cross-fertilising ideas between functions; and at collaborating across functions.

Get your engineering department to run your background product development programme.

Function or departmental-based organisations are better at developing functional skills, new technologies and keeping up to date; better component performance results. They are most effective at developing standardised and optimised individual components, building blocks and systems. You can get your engineering department to run your background product development programme. Their specialist expertise will arrive at the best result. But departments are not good at organising projects, multiple or single.

When running a number of projects, dedicated product teams will complete the task in the least elapsed time but can use more engineering hours. They can also make task sharing between multiple projects more difficult. To avoid this, you make the teams share the responsibility of transferring technology one to another.

Another solution is to use functional departments and project teams with a layer above to co-ordinate across the projects (see section 10b.6). The best solution appears to be continual liaison between project managers; this gives direct co-ordination between equals. You direct the multiple projects by the project managers and centre manager together.

What you do is driven by your strategy. Do you want efficiency rather than simpler project management? And how should you group your products, and into how many centres? Just grouping them without logic is not effective. Should the grouping be by market or technical similarities? Using common platforms as the logic works well. You also need supporting mechanisms and processes for multi-project management. And you need creative engineers and product champions to produce successful products.

10b.4 The dangers of bottlenecks

When several projects compete for facilities, such as specialist simulation or testing, you need more than the 100% capacity that you believe will be sufficient. Lack of capacity will severely delay your programmes. And that will cost you far more than installing more capacity.

If you are calculating capacity accurately to make sure the facility is used 100% of the time in order to save cost and get approval for the capital expenditure, beware. Calculating capacity at 100% utilisation will cost you dearly. Why? Because it's impossible to schedule multiple projects so precisely that none ever over-runs any task in its programme. You just can't schedule with 100% accuracy to utilise a specialist facility 100% of its time. As soon as one programme departs from its ideal timing even slightly, that will create a queue in front of the facility. All the programmes will thereby be delayed. The cost of this usually makes the cost of increasing the capacity look insignificant. The problem is that the true cost of delays is not calculated and therefore delays are accepted rather than their cost being highlighted. So the best economic action becomes hidden from general view. Reinertsen[6] gives good detail on this aspect.

10b.5 Component strategy and background development

You do better if you can apply key components across multiple product lines while the technology is relatively new. Speed is the key. Relatively few companies take advantage of this concurrent technology transfer.

> You also need good planning to ensure that the new and shared building blocks are on time for the cluster of new projects. You will need to manage suppliers' programmes as a part of your in-house programme.

But to get innovative new designs, you first have to develop your new technologies – continuously as background. And this background continuous development may be better done by the relevant functional or component department.

10b.6 An operating example

A successful way is the differential matrix type of organisation. A good large-company example is how Toyota did it; Mazda adopted it later. They did the following:

- brought technically similar projects together into centres;
- maintained a consistent design philosophy for their products;

[6] *Managing the Design Factory, Chapter 3,* Reinertsen.

- reduced the number of engineering functions in each project; and
- created a new centre for common component systems.

This reduced the size of each centre manager's management task so they could cope with overseeing all the activities within each centre. And they gave each centre a planning section to support its work. They also created a specific development goal for each centre to focus their efforts. The centres competed against each other then shared what they learnt. This accelerated the rate of improvement and of generating new ideas.

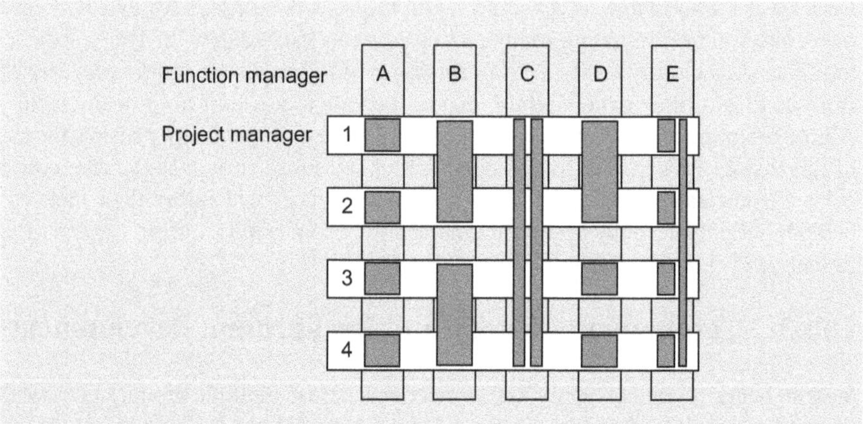

The figure shows the principle of how a centre was organised. A centre manager managed the whole:

One project manager takes the lead in joint projects. For example, in Microsoft Office, the project manager over the Microsoft Word team takes the lead in developing text solutions across all programs. Each engineer has a dual responsibility, for example for project work; or for particular components across 2–4 projects (e.g. fonts).

Smaller companies may not have separate centres but they still need the supporting mechanisms and processes for multi-project management. They also need to solve how to integrate different engineering functions to create distinctive individual products while sharing technologies and co-ordinating multiple projects.

10b.7 Implications for staffing

People should stay together for succeeding project generations, so they learn and carry a consistent design philosophy from one generation to the next. This works better than disbanding and creating a new team because face-to-face contact is more effective in transferring knowledge.

Project managers and centre heads should be explicitly established posts. And you have to plan the authority levels of project managers and centre heads to make sure that they and the department heads have clearly separated responsibilities. If you don't, you will suffer frequent organisation changes to cope with the problems. You need stability and consistent focus because developing successful products is difficult, costly and time hungry. If your staff are uncertain of their roles and responsibilities they will waste time working out who is responsible for what and to what extent.

10b.8 Further detail

The information in this chapter is mainly derived from the references cited in *Chapter 12*. Refer to them for greater detail.

Part c Risk assessment

Quantify the risks that are inherent in your proposed programme; reduce them as you proceed.

10c.1 Scope and application

Risk is defined as the probability of failure multiplied by consequence of failure. You manage risk down by applying resource to reduce the probability of failure, especially of high-consequence events.

Risk assessment quantifies the risks inherent in a programme, and allows the precise nature and elements of the risk to be identified. The underlying purpose of managing a product development project is to reduce the risk of failure progressively, or terminate when the risk is recognised as unacceptable.

10c.1.1 Relationship between risk, project size and number of phases and reviews

Even programmes with low capital or development cost but indicated high risk need rigorous project management to make them successful.

> This section shows a very simple comparative risk-assessment method. If you want a full-blown complex assessment of an exceptional project, you can expand each category into more levels of detail. Although it may seem overly simple, it provides a sound foundation.

For programmes in the low-risk category, merging phases where appropriate can reduce the number of reviews. When phases are merged, merge their transfer criteria but do not omit them.

Conversely, high-risk programmes, irrespective of size, may need intermediate reviews at the pre-authorisation stages, to ensure control and elimination of the factors giving rise to risk. This is at the project champion and review team's discretion.

10c.1.2 Process

Using internal company knowledge and judgement, a cross-function review group does a fast outline risk assessment by the process described below. Each member does their own. The results are then collated and the group meets to discuss the result. The purpose is to arrive at common understanding (a revelation to most companies). This discussion alone will revolutionise what you include or omit from your programme.

The risk assessment is reviewed as part of each successive project phase. In Phase 1 the project mentor arranges this; for subsequent phases the project manager or leader does.

Once high-risk elements are identified, you devote effort to managing them down to an acceptable level. Usually, the higher the identified risk level, the more reviews and resources will be needed.

> Keep the method extremely simple or people won't find time to do it. Use it at board level to make sure everyone has the same understanding of the project. And use it to prioritise.

10c.2 Risk assessment

Assess three principal risks:

1. **Technical risk** – for example, the risk that the programme will not produce the right performance or quality of product or that other customer requirements may not be met. The programme must not expose the public, employees or the environment to the unacceptable consequences. The newer and more novel the product and process is to the company, the greater are the risks.
2. **Commercial risk** – for example, the risk that the output cannot be sold because of aggressive competition or lower than expected customer demand, or wrong marketing strategy or because the customer requirements have changed during the project. An example is European consumers' resistance to genetically modified foods.
3. **Financial risk** – for example, that the project will not attract the necessary capital or will fail to generate the expected returns because of, say, incorrect costings, currency movements or failure by third parties to meet commitments.

10c.3 Quantification

Quantify project risk under headings of Technical, Commercial and Financial, using objective or subjective views. Score each item 0, 1 or 2 depending on your assessment. Where an item does not apply or shows no risk, score 0; for intermediate risk score 1; for the highest risk, score 2.

Tailor this basic to your own products but limit each list to around 10 items.

Technical risk

Risk description	Score 0, 1 or 2
Product novelty	existing product/new variant/new product
Product technology	proven–speculative
Number of interfaces	low–high
Process technology – manufacture	well known/simple–less known/complex
Process technology – assembly	well known/simple–less known/complex
Synergy with product strategy	high–low
Certification approvals	none required–partial or full
Environmental health and safety problems	none–major
Peoples/facility resource	available–major new requirement
Site operational experience	familiar–new
Ownership/control of PD process	shared–not shared
Joint ventures – suppliers and partners	
Partners' product technical strength	strong–medium–weak

Commercial risk

Risk description	Score 0, 1 or 2
Strategic fit	good–poor
Market location	existing–new
Market knowledge	familiar–poorly understood
Route to market	existing–new
Customer base	existing–new
Customer/dealer acceptance	good–poor
Market share for success	small–large
Service requirement	low–high
Urgency (high = highest value of project)	high–low

Financial risk

Risk description	Score 0, 1 or 2
Capital at risk(set your own values)	low (e.g. <\$20k)–high (e.g. >\$100k)
Affordability/justification	easy–difficult
Cash flow	easy, moderate, hard to fund
Payback time (set your own values)	short (0–1.5 years)–long (>3 years)
Sensitivity to volume	low–high
Sensitivity to (full) costs	low–high
Affordability for customers	easy–difficult
Currency exchange rate	sensitivity: low–high
Political sensitivity, internal and external	low–high
Plant location	well-supported local–remote
Joint ventures – suppliers and partners	
Partners' financial strength	strong–medium–weak
Partners' knowledge of business	high–medium–low

10c.3.1 Explanation of terms

To train staff, first define what each category includes. Don't allow nit-picking to prolong the analysis: it's supposed to be a simple overview so that everyone can share their perception of how risky the new project may be. The aim is to make everyone share the same view of the project.

Technical risk

Risk description	Explanation
Product novelty	Is the proposed product well known, or unfamiliar (novel, innovative)?
Product technology	Is the proposed product's operating technology well known, or unfamiliar/novel/innovative?
Manufacture technology and assembly technology	Is the proposed manufacture/assembly technology in the plant where it will be made well known, or unfamiliar/novel/innovative)?
Certification approvals	How difficult might certification approvals be?
Environmental health/safety	Might the product or its production process harm anyone?
Peoples/facility resource	People and facilities, not money – money is dealt with below.
Site operational experience	How experienced is the company at making this kind of product?
Ownership of process	Is ownership of manufacture shared or not?
Partners/suppliers	Confidence in their technical competence

Commercial risk

Risk description	Explanation
Strategic fit	Does the product fit the company's product strategy?
Market location	In which markets are you intending to sell the product? (existing–new)
Market knowledge	How well do you understand those markets? (well–poorly)
Customer base	Is it to be sold to a company customer base that is already established?
Route to market	Do you intend to use an already established route to market?
Customer/dealer acceptance	Will dealers and customers like the product and want to buy it?
Market share for success	What level of market share do you need to be commercially viable?
Service requirement	What level of customer service costs will you have to provide?
Urgency	How soon must you introduce the product: will it need to leapfrog others in the programme? Rank in order of financial value of completed project.
Partners/suppliers	Confidence in their commercial competence

Financial risk

Risk description	Explanation
Capital at risk	For example, low risk might be <£20k, medium £20–100k, high >£100k
Affordability	This is related to strategy: if you must have it, it's usually 'affordable'!
Cash flow	Can the company fund it now (easy, moderate, hard)?
Payback time	For example, short might be <1.5 years, medium 1.5–3 years, long>3 years
Sensitivity to volume	To what extent will low sales volume render it financially unviable?
Sensitivity to (full) costs	To what extent must you recover the full costs of development and manufacture?
Affordability for customers	How affordable will it be to customers (easy, moderate, hard)?
Plant location	Might the production facility's location cause extra problems and costs?
Currency exchange rate sensitivity	Could the currency exchange rate of proposed markets increase or reduce sales income?
Political sensitivity	Might the internal/competitor/national political situation cause problems, where either manufacture plant or customers are located?

10c.4 Computing programme risk *(sum of individual risks)*

The total risk is the sum of the individual risks. Don't complicate matters by multiplying the individual risks together to arrive at the total. Just sum them: this is not an academic statistical exercise but a simple management tool that everyone is trained to use.

The risk thresholds below are for guidance and you should check them against your past project experience. But note the caveats:

- apparent low risk can escalate through neglect and can creep into other areas if not identified and managed;
- a score of 10, comprising five 2-score items, is high risk; whereas 10 comprising1-score items is low;
- take special steps to manage 2-score risks down early or they may delay or stop the programme later; for example form a special team; or build and test prototypes very early in Phase 3 to make certain they work and will be reliable.

10c.4.1 Very low risk

Programmes with a total score below 5 that do not have any items that score 2. If this is combined with fast payback, you could consider fast-tracking the programme.

10c.4.2 Low risk

Programmes with a total score below 10 that do not contain any 2-scores have relatively low risk. The items with a score of 1 must be managed down before launch.

10c.4.3 High risk

Programmes with any 2-scores need special management. Do sufficient work during the concept phases to reduce the 2-scores to 1 or 0. Do not proceed beyond the concept phase if you can't do this. You can put high-risk technical aspects into your long-term product development programme and put them into later projects when their risk is acceptable.

10c.4.4 Unacceptable risk

With large numbers of 2-scores, either shelve the project or modify the highest risk aspects to reduce the total to an acceptable level. But if the potential reward is huge, do *Phases 1 and 2* before you cancel.

Note: Time is a form of risk

Time may be your most critical risk factor. The longer you take, the more risk you add.

Beware that products fail mainly because customers reject them. Companies seem less able to please customers than to solve technical problems. Poorly (or not) researching their needs is a major cause. But a major element is the time factor. The longer you take to develop a product, the more product requirements will change because customers develop their tastes and change their minds. Customers' tastes can change faster than a bureaucratic company makes decisions. Some companies vacillate so long that customers change their tastes beyond the product even before its development is approved.

10c.5 Further information

Reinertsen gives more extensive background detail on risk management.[7]

[7] *Managing the Design Factory*, Reinertsen.

Chapter 11
Tools and techniques

It's not enough just to follow a logical process. You use tools, techniques and methodologies to ensure that each task is done as effectively and completely as possible. Model the economics of what you are doing, so you know how to control the right things in the right way.

11.1 Introduction

11.1.1 The tools and techniques that help you cut time and cost – what they enable you to do

Tools and techniques are useful to reduce time and at the same time permit increased product complexity. But don't learn how to apply them in detail until you are about to use them. It's like learning a foreign language at school then needing to use it 20 years later; as well as forgetting, you didn't realise what would be useful or useless!

There is an armoury of tools and techniques. But confusion exists because there are so many, with different and sometimes overlapping functions. Some can be used in more than one phase of the process.

The following brief catalogue describes what the more useful ones can do. A fuller list appears in John Bicheno's book. The aim here is to describe not how you operate them but what they can help you to do. A number of works, referred to where appropriate, describe how you can learn to use and apply them in your programmes; but only when you are ready to.

11.1.2 Contents of the chapter

The logical order in which tools should be listed is arguable – you are bound to prefer an alternative! If you do it alphabetically, you jumble the subjects. If you do it by subject, you can debate which really fits where; it's easy to re-classify them a number of ways.

The order in this chapter is related to my experience of the order in which it's useful to learn about them. For example, QFD can require a fundamental re-thinking in the way you organise your staff to communicate throughout the company. So to an extent does concurrent engineering. The two can go together. But if you understand the reasons for using QFD first, you may re-think the way you are going to organise your concurrent engineering. You may organise it to encompass more areas that you might have done before you had read about QFD. And both require team working, so it's logical to deal with that after both. Then incremental innovation is more than a design process; it relates to strategy. So it's not really appropriate to include it under prototyping and designing, although that is where the end result is put into action. (You can have your own debate about the order in which they should appear.)

11.2 Quality function deployment

11.2.1 Origins

The words *Quality function deployment* are a literal translation from three Japanese characters. They don't translate well into English. It was intended to mean 'ensuring that the (qualities) performance and specification demanded by the customer of the product are analysed and prioritised then clearly (deployed) communicated to all (functions) areas of the company'.

QFD is one of the most valuable techniques to help you develop products that will be really successful. It's relatively new – devised by a Japanese academic in 1963, first applied in a Japanese shipyard in the late 1960s, in some large US corporations in the late 1970s, and 'discovered' by Europe in the 1980s. But so far it's not used much in Europe because, although a minority has heard of the name, few understand its immense benefits. By 2000, fewer than 50 UK companies were trying to use it. It's not easy to apply, but disproportionately rewarding when mastered.

11.2.2 Introduction

QFD is a formalised methodology of making sure you design and develop your product to give customers exactly *and only* what they want rather than what someone else believes they might want. It's a formal process to avoid your wasting money by giving them what they won't value and don't want to pay for. It's all about giving the customer the best value.

Features that customers don't want often originate from what best suits company departments' internal ease of working, or from peoples' pet views and theories on what they think the customer *ought* to want.

QFD forces everyone to listen to the real voice of the customer and to focus on real requirements. It prevents wasting resource developing features that customers don't want and are reluctant to pay for. If you develop features that customers don't value, you alienate them because they think it costs them more. In such cases, you may have to give the features away in order to get them to buy the product at all.

QFD ensures that the product performance and specification demanded by the customer is transparent and clearly communicated to all areas of the company.

11.2.3 The benefits

Every chief executive would benefit from an outline understanding for two fundamental reasons. First because it minimises what you need to spend in order to give your customers exactly and only what they want. You concentrate your resource on analysing what really matters to the customer, to ensure that the product incorporates all the features they want, and nothing they do not value. Its objective is to avoid wasting company resource on features the customer does not value: you minimise your development time and your manufacture cost. It reveals flaws in judgements at the earliest stages so you develop better concepts and direct your effort to best effect.

Second because it brings teams together to gain a common understanding of all the issues. It encourages a unifying team-approach across the entire company in a real working project, and gets fast results. It is not just an exercise.

It is a vital part of the product definition stage. It makes sure you translate the well-researched features through every stage of the product's development. It fits into the following sequence of activities:

- setting company objectives;
- researching information on customers;
- QFD stage one;
- setting the first requirement for the product, the marketing specification.

> The QFD process starts by formally analysing what the customer wants, and then compares that with the features responsible for product performance.

It dovetails into concurrent engineering and assists facilitate several other tools and techniques. You need perseverance to learn and apply it. It needs a fairly rigorous mental discipline and attention to detail. But the effort is outstandingly profitable. You gain a competitive advantage over those who don't use it either because they don't understand what it can do or because it looks daunting. Those who persevere gain a big advantage.

11.2.4 The process

The QFD process starts by formally analysing what the customer wants. It sets out in priority order the characteristics that they value most. Do you know what they are for your product? Have you formally researched them? Most companies never have; they tend to look at competitors and their own opinions on how they would use the product.

You then list the product's physical parameters that satisfy each customer need. Measurement tools prioritise the importance of the product's physical features to the user. It gives a tool to arrive at the best trade-off between features where some reduce the benefits of others. It extends to comparing the performance of competitors' products with that of your own.

What should you do if some of the product's features conflict? Suppose the market is so fragmented that no single segment appears to be viable on its own? QFD addresses this by examining the relative value that potential customers place on the possible product features. You use the information to optimise the specification, maximise appeal across a number of markets, or maximise your margins or your market share in one particular market – whatever your objectives are.

No process that measures such attributes can yield absolute values. But they can reduce risk and help ensure that a larger proportion of your potential customers will want your products, and that you will discover what variants are needed and for which markets. Your team collects information on

- the relative value potential customers attribute to each feature, including price sensitivity;
- how they would trade one feature off against another;
- how customers from different market segments differ in their perceptions.

You run QFD in stages. Each stage is between two interfaces of the product development process: stage one is the customer–company interface, used to validate the detail you need for concept design. Stage two is between concept and detail design ... and so on. In effect you start by running the first stage then reviewing the specification and achievement at the end of each stage and comparing it with your original QFD matrix.

Every company should use it, if not in full, certainly as a routine part of developing its products. Clausing[1] gives guidance on how you run the process, with examples.

11.3 Concurrent (simultaneous) engineering

> Concurrent engineering will work correctly only if you use a team drawn from all areas.

Simultaneous engineering is the quickest way to arrive at the best product solution. First, you gather the wisdom and expertise of people from all fields, at the earliest possible stage. The process will work correctly only if you use a team drawn from all areas. If you don't, you will not gather enough information at the outset and that will cause too many changes later in the programme – a recipe for increasing time and cost that is avoidable.

Second, you complete tasks in parallel rather than in series: you start each task with the minimum possible information rather than waiting for the previous stage to complete its task. You complete the programme in far less cost and time. And you get better ideas that improve the product both for the customer and for you to manufacture.

[1] *Total Quality Management*, Clausing.

Some companies have more than halved the cost and time of developing new products. To operate effectively the process needs good initial planning as outlined in section 9.1. It fits logically with QFD (see section 11.2). And it cannot be run without team working (see section 11.4).

When experienced and competent, you will need far less overall resource. You put more emphasis and resource into the initial stages, which speeds progress. You will have a smoother introduction with fewer changes. You will foresee production and customer problems that you never did before.

Initially companies need to gain experience and learn the rules. There is a working manual[2] that enables you to put concurrent engineering and team working in place, step by step. It draws on the experience and lessons learned by companies that have successfully done it. You can manage the change with as little risk as possible. It takes your management team through the whole process in a structured manner. The manual sets out how to plan and run in-company sessions. A number of companies have used it with excellent results.

Companies that do it for the first time often encounter challenges they were not prepared for. In some cases, this has caused them to abandon the process. You may encounter problems if this is your first time; but with experience and foresight, you avoid them. Typical areas are

Structure and organisation

- Poor product development capability;
- Poor project management and control;
- Slow flow of poor information;
- Poor relations with your customers and suppliers.

People

- Need for a team drawn from all areas with a flexible attitude and a commitment to improve continuously;
- Need for staff with a combination of commercial and technical skills;
- The ability to work with partial information and cope with uncertainty.

Funding

- Higher level of resourcing of initial stages;
- Inevitable initial costs of errors because of new ways of working;
- Incompatible payment terms between suppliers and sub-contractors;
- Less day-to-day financial control by the highest level of management.

You need to consider three levels of concurrency:

1. Project concurrency, overlapping product development activities.
2. Design concurrency, overlapping the design disciplines.
3. Product concurrency, running multiple projects (see *Chapter 10, part b*).

These aspects are covered elsewhere in the manual.

[2] *Using Concurrent Engineering for Better Product Development,* Cranfield University.

11.4 Team working

In developing products, using a team that represents all areas of your company will reduce your time-scales and, in the long term, your costs. Most world players use it in conjunction with concurrent engineering. Together, the two can achieve 80% of the possible cost and time savings.

Teams ensure the best and widest possible input of information. This does not just involve engineering departments or engineering activities. It focuses on collecting information from all company facilities, personnel and departments from market research through all areas the product touches on its way to the customer. It also includes suppliers that are developing their components or products that you will use in yours. A multi-discipline team creates an automatic exchange of information between functions and firms. It can reduce 80% of late engineering changes by avoiding the need to re-do tasks through discovering too late what the requirements really are.

Some suggest that companies will benefit from moving to a totally project-based structure with co-located multi-discipline teams. Here, the core departments shrink to small excellence cells that receive team members between projects and tutor them. *Chapter 5* explains this in more detail. But culture is more important than location or organisation – it's culture that makes it all work.

When moving to team working, note army experience. Officers are taught two ways of commanding a brigade or division in the field. One is Directive Control (DC), which means laying down absolutely clear general orders and priorities and then letting junior commanders carry them out. Whereas in Order Command (sometimes called leading from the front or, less kindly, obsessive interference in detail), a general ensures that his operational orders are being carried out precisely, and jumps in at any moment to overrule orders given by a subordinate.

The most effective companies use the DC analogy and have a collaborative culture between mother departments and teams. Toyota, for example use a matrix organisation with such a culture, and do not always co-locate their teams. It's their culture that makes it work so well, together with their managers' remit to continually tutor their staff. See section 5.12 for more.

Many companies find it difficult to put this in place because of entrenched management attitudes. The Cranfield University reference in *Chapter 12* is an excellent practical guide on how to overcome this.

Functional department heads often have a 'vertical silo' mentality and see losing management control of their staff as a threat to their authority. They don't understand the benefit that team working brings to the company's profitability, or to expanding the experience and expertise of their staff. Their entrenched view is born of a perceived insecurity to their own future rather than being viewed as a way to expand their sphere of influence on the company's future.

Good engineering requires intense communication, which in turn requires organisation. Engineers need to spend time in the workshop, in the office and in meetings. Meetings can be a great source of waste without discipline agendas, and with no one writing minutes. Communication can be enhanced by co-location of project teams and through the use of tools like forms, one-page reports, copy boards and on-line discussion databases.

11.4.1 Company culture

But how well your teams operate does not depend on whether you choose to work in vertical silos, a matrix or a project-based organisation. What governs performance is company culture: do people work well together across department boundaries naturally, without conflict or turf wars? If they don't, no amount of organisational change will get your teams working. It may make you feel better if you reorganise; but it won't make your teams work if the culture is against it. It is ***The*** key 'soft' issue that makes or breaks a company's product development capability. If the CEO doesn't support and nurture such a culture, you won't have an effective capability. See also section 5.12.

> There's no point in having a marvellous process; if you have the wrong culture, the process won't work.

11.5 Incremental innovation

Incremental innovation is where you improve products in frequent small steps rather than in sporadic epoch-making leaps. It greatly reduces the cost and risk of introducing new products – you can do it quickly and at low cost to maintain or increase market share. At the same time it reduces the time and cost of developing products because you get more practice at running the product development process.

If you are developing a variant of an existing product, do not make small incremental changes to a lot of areas, whether they are visible or invisible to the customer. Because they are unlikely to notice them and you will not recoup your investment.

> Incremental innovation introduces a really large change to not more than two or three features that will be very visible to the customer.

Instead, introduce a really large change to not more than two or three features that will be very visible to the customer, which they will really value. This will create a product that is obviously different from its predecessor that you can promote as 'new'. To everything else in the product, make small incremental improvements (maybe 5%), to improve aspects such as reliability, and always to reduce cost and to ease manufacture and assembly. By this method, you introduce new products quicker at less cost and risk with faster market success and return on investment.

The aim is to ensure that each new model has even more customer appeal. It generates a steady stream of new products that keeps your competitors on the run, preferably behind you trying to catch up. If you are behind them, you can use the process to catch up, faster and at less risk and cost.

11.5.1 Radical innovation versus incremental

Innovation is related to how old or new the new product is compared to your previous ones. Companies have a better chance of success if they develop products based on technologies or markets related to previous products. This works better than products that are totally unrelated to existing products or are intended for totally new sets of customers.

But you are unlikely to remain competitive by only following incremental product innovation. You also need to introduce perceived newness through technology or design features. You need to combine the incremental with the innovative. But this must not be so radical that takes too long to become accepted to attain your sales targets because it alienates customers. Or take too long to develop and as a result cost too much. But realistically, you can't produce radical innovations continuously. You need another factor – the advantage you gain from your ability to organise projects rapidly and bring them to completion faster than the competition.

Incremental changes may need no more than reinforcing the discipline to follow existing processes. Whereas to develop truly innovative (to you) products, you may need to change your existing organisational capability or even set up a separate organisation.

You need to use your accumulated capabilities and resources across your whole product range. Your core competencies are not just innovative technology but the ability to utilise it in multiple projects. You need to optimise the attractiveness of your product portfolio as a whole. For example, you can use architectural innovation – the way you combine the systems within each product that you are developing. Refer to *Chapter 10, part b.*

11.6 Brainstorming – generating ideas

Brainstorming is used to generate ideas. Why use brainstorming? Because

- it collects a large number of ideas very quickly;
- it enables you to identify common groups of ideas and patterns;
- it encourages everyone to participate; otherwise the loudest voice might predominate and result in not collecting everyone's ideas – you want all ideas; and no one must feel left out;
- it records information for later use.

Brainstorming avoids the loudest voice predominating and collects everyone's ideas very quickly.

You run the brainstorming sessions first by a facilitator explaining the aim of the session. Then they present a question you want everyone to brainstorm. Everyone answers the question, not what may be behind it; they don't interpret it, manipulate it or try to be political.

They focus on the question in silence, then using a *post-it* pad, write one idea per *post-it*. Freewheel; don't evaluate. Record as many ideas as possible; don't worry about the quality – go for quantity. You will sort quality and evaluate ideas later. And write large so everyone can read it from three paces away; express the idea in the minimum number of words.

Everyone then sticks their *post-it* notes on a wall chart and sorts and groups the ideas. You gain a large number of ideas and everyone participates. A fast effective process gives a fast, constructive outcome. And it helps generate trust between the participants. The result is a series of grouped ideas, which you can then evaluate in detail to arrive at one or two that are feasible to take forward.

11.7 Pugh concept selection

Selecting the best concept for company and customer from a number that have been devised is not lightly done. There is an excellent formal technique for doing so however. The late Professor Stuart Pugh of Strathclyde University devised it over years of successful practice.

Its great advantage is that having developed a number of alternative concepts; the technique results in the team devising new improved variants. The one finally selected is invariably better than those previously devised. Although it is a selection process, it almost always results in a new concept that did not exist previously.

Pugh concept selection results in the team devising new improved variants. The one finally selected is invariably better than those previously devised.

You can use it to develop concepts for system architecture, sub-systems or individual components. It works best when the production method is considered with the design.

The method selects the criteria that are important to the customer and company, and forms an evaluation matrix. It then compares how well each concept satisfies the requirements compared to a chosen datum. From evaluating the ratings and enhancing the positive features, new concepts result and arrive at a winning concept.

The starting point is where a number of concepts will have been devised and checked against criteria of customer needs and company strategy. The marketing specification criteria are entered as column headers in a matrix. Each concept is a row in the matrix. Selecting one concept as the datum, each concept is scored as *plus*, *minus* or *the same* against the datum for each of the criteria.

The scoring is not straightforward. Initial perceptions change and the resulting discussion develop new insight. New hybrid concepts that were not previously considered will emerge. At the conclusion of the discussions, each concept will have a score. The best is taken forward. But awarding scores is only part of the methodology: equally important is the insight the technique generates to devise new concepts.

From the initial exercise, many of the original concepts will be ruled out and new ones added. The team will now have a better understanding of the issues. The exercise is therefore repeated, using a new datum. This results in selecting a dominant concept that will usually be a new one developed from running the technique. The one chosen will undoubtedly be better than the ones originally brought into the exercise.

The technique avoids rework later in the programme that might otherwise have occurred from new ideas creeping in. The technique is described with examples in more detail in Clausing's book[3] on product development.

11.8 Functional cost analysis

This is a technique for analysing what it costs to provide each function within the product. It's best to use the functional cost analysis in conjunction with optimising the concept. You can apply the technique to your completed concept design and devise a matrix with a component part on each row and the desired functions as columns. The functions are derived from your QFD analysis (see section 11.2). On each row you allocate the proportion of the component's cost that contributes to each function. By summing the columns you arrive at the total cost of providing the function. The costs of each thereby become visible so you can address anomalies. Clausing[3] expands on the detail. It is related to value analysis in its effect but the formal methodology may be easier and faster to master by engineers with analytical minds.

11.9 Failure mode and effects analysis (FMEA)

Originally developed in the aerospace sector, FMEA is a systematic planning process, which identifies problems that could occur. It examines the function of the product and identifies potential sources and mechanisms of failure and the effect this would have on the product's function.

The technique is useful not only during designing but also in reducing problems during manufacture and installation; it can produce huge cost savings, especially from reduction in failures under warranty.

[3] *Total Quality Development*, Don Clausing.

FMEA is a systematic planning process, which identifies problems that could occur.

You examine all the possible ways that a product or process might fail and devise counter measures for each. You can do it on any product – for example, on how a component or system might fail in service (design FMEA), or how your customer might (mis-) use your product.

You can also apply it to a production process or to a business process. See also section 11.12 and *Chapter 10, part c, Risk Analysis*.

It is a systematic, element-by-element, analysis to discover where failure is possible. Every conceivable mode of failure is schemed out and the potential reasons catalogued for each. You then find out what is currently done to detect each cause of failure and introduce corrective controls. You design the controls to detect whether the corrective measure has been employed or not. This ensures that the failure does not occur.

11.10 Cause and effect analysis

Ishikawa is a technique that considers all the possible causes of a problem and places them in a logical order for investigation. It forces a full analysis of a problem to ensure that every facet and avenue of it is properly explored and understood. Its use can avoid blind alley investigations that are working on areas, which are just not relevant to the problem in hand. It leads to the significant being identified from the myriad of insignificant factors.

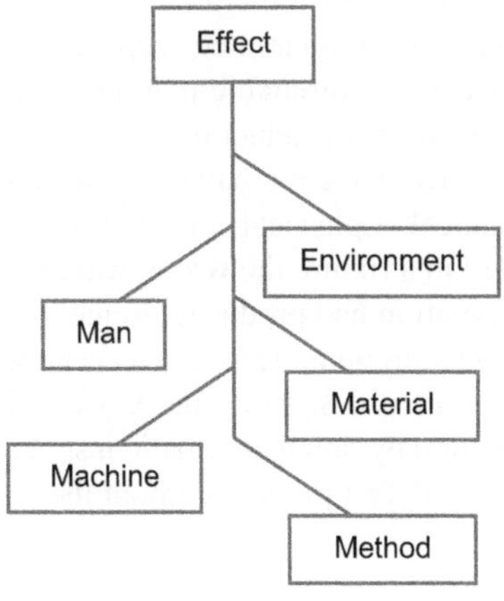

The fishbone diagram on the left is an illustration of how the technique can be applied. The problem may be caused by a number of elements – man, material and so on. Each possible cause is examined, which may result in further subdivisions branching off these main headings. It provides a structured logic to examine the problem and its causes.

11.11 Five whys

This is a method of trying to identify the root cause of a problem by repeatedly asking the question *'Why?'* until you can't ask it any more. This may be more or less than five times, but most problems yield a root answer at five or less.

For example, why did the product fail? Because an electric motor failed in one of the system modules. Why did the electric motor fail? Because it overheated. Why did it overheat? Because the air around it was too hot. Why did the air around it get too hot? Because there is no ventilation. Why is there no ventilation? Because the housing air outlets are below the level of the motor. Why? Because this module was changed and the outlets are positioned for the previous module. (So either alter the housing or the position of the motor.)

11.12 Pokayoke

'I was shown many Pokayoke devices invented by the employees to prevent defects. Beside each Pokayoke was a card explaining its purpose and who had invented it.' – visitor to Matsushita washing machine plant in Shizuoku.

Pokayoke is mistake proofing, a preventive measure. It is a simple, often inexpensive, production device that prevents defective parts from being made by sensing abnormalities. It does not rely on human action or memory. Its effect is 100% automatic inspection. You foresee problems and devise a means to prevent their happening, rather than wasting time and money by trying to discover them later by having to inspect them. It guarantees zero defects through the production process rather than by relying on unreliable judgmental inspection.

It usually takes the form of a production jig or fixture. It prevents defective operations that would produce defective parts. At the same time it achieves 100% inspection through mechanical or physical control. For example, it might check that a part could not be inserted in a fixture the wrong way; or that the part contained all the features a previous operation had produced, to identify where some might have been omitted. It's a production-engineered intervention that avoids producing defects. You devise them when designing the part and its production method. It ensures zero defect production by doing automatic inspection. If you do it correctly, you therefore achieve a zero-defect process without the need for human inspection. Shingo[4] explains the detail.

[4] *A Study of the Toyota Production System*, Shingo.

11.13 Weibull analysis

Weibull analysis is used to analyse, for example, the pattern of failures in service. It is based on a statistical methodology and can separate different failure modes out of an apparently random pattern of failures. There may not be public training courses on the subject – large engineering companies (especially mechanical or consumer) tend to use it as a routine tool. Contact the engineering department at your local university for assistance, or for a reference list on Weibull; it will be part of their engineering degree syllabus. *Practical Reliability Analysis* by O'Connor gives detail on the subject.

11.14 Process capability

Process capability is measuring the statistical capability of a machine or process. The process is statistical common sense. It needs to be done before you can do SPC (see section 11.15).

When a machine is set up to produce a series of parts (or other output such as dispensing volume), the apparently identical parts vary slightly, one from the next. If you plot dimension against frequency of occurrence, you get a normal distribution; process capability is defined as six standard deviations (6σ). It's the dimensional spread within which 99.5% of the output lies.

You use process capability when you specify a tolerance to a dimension – linear, volume, weight – in a manufacturing instruction. Before you can specify tolerance as a dimension, you need to know that it's possible to achieve – which requires knowing the process capability of the machine that will be used. If you don't have this information, you can specify a tolerance that is too tight and quite unrealistic. This results in ridiculously high reject rates and causes many a quality moan. See section 11.15, on how to do it.

11.15 Statistical process control

SPC is a technique to ensure that components produced in quantity from reasonably complex processes are all within tolerance without having to measure every single component.

You measure the process capability (6σ – see section 11.14) then set the drawing tolerance to $\pm 3\sigma$ for the dimension you're specifying. Having measured the process capability of the machine, you know that it can reliably produce output that will lie within that tolerance. On the machine, you set up a process range chart showing the $\pm 3\sigma$ tolerance versus number of parts produced, and you add a $\pm\sigma$ band within that. You adjust the machine only if the output veers repeatedly beyond $\pm\sigma$. You check one in so many, not everyone. If the measured dimension is inside $\pm\sigma$, you know that (barring catastrophe) they will all be within tolerance. To avoid out-of-tolerance product, you need to adjust the process only when the measurement starts to trend towards one edge of the $\pm\sigma$ band (from tool wear, for example). It's a sure method of reducing defect levels to near zero at little on-cost.

11.15.1 Six Sigma

Six Sigma aims to identify and remove the causes of defects, to minimise variability. It uses statistical methods (see section 11.14 and 11.15) and follows a defined sequence of steps with quantified targets. It was originally developed in 1986 by Motorola. The internet bears copious information.

11.16 Taguchi methodology and robust design

Professor Taguchi is best known for his work in devising designed experiments to cut the time needed to diagnose problems where multiple variables interact. But Taguchi methodology is concerned with far more: reducing quality loss.

The aim is to ensure that the design process enables all finished products to produce the performance that customers want. It does this by focusing on robust performance. You find the design parameters that are critical to the product's performance and control them.

The aim is to improve customer satisfaction, reduce development time and cost. Optimising robustness as early as possible in the product development process reduces later rework and therefore total time and cost.

This is not the same as a zero-defect policy. You could produce zero defects to the tolerance levels you have set. But your tolerance levels could be wrong – some too wide, most too narrow. So the product would vary widely in its performance even when they were all within your tolerances.

The aim is to reduce variations from the product's ideal performance. It is done in four stages:

1. Systematically find the particular design parameters that are critical to achieving the performance;
2. Set the economic optimum for each key parameter by tolerance design;
3. Set economic levels of the most important processes that are needed to achieve the required accuracy and repeatability of the key tolerances;
4. set the on-line control levels needed.

The key activity is first – identifying the key parameters to control. Second, finds the level and tolerance of each parameter that will keep the performance as close as possible to the designed performance. The others ensure its achievement.

You design experiments to find out what critical factors you must control to produce the designed performance at minimum cost. This results in so-called robust design.

11.16.1 Robust performance and designed experiments

Lack of robustness results in many, or even all, of your products not reaching the performance you intended them to achieve. The performance of a product will vary for three principal reasons:

1. A product can perform differently depending on how and in what conditions the customer uses it; robust designs perform well in all conditions because tolerances will have been set for the parameters that are key to its performance in all conditions.
2. Variations in how the product is made will cause its performance to vary and some products will be unacceptable to the customer; a robust design will have identified the level of control needed in the critical processes to avoid this.
3. Products will deteriorate in use, for example from wear and tear; the extent to which they do can be designed to limit performance degradation.

Lack of robustness will result in many, or even all, of your products not attaining the performance you intended them to achieve.

For example, a water meter manufacturer designed a component that had to be watertight and had certain dimensions that were critical to its performance. It was designed to be made from ebonite. But 80% of the mouldings were unusable. They were outside the critical dimensional tolerance, and even worse, most were very porous. There were six process parameters that could cause distortion and porosity. Controlled experiments diagnosed that two key parameters interacted to produce virtually the whole effect. Both needed precise control – temperature and its time and uniformity when preheating the moulding blanks, and the tolerance on the time and temperature of the moulding process. The other four parameters were not critical and did not need such precise control. Experiments verified the tolerance bands needed for both critical processes. Measures were put in place to ensure that these were controlled automatically. The resultant process was robust enough to reduce defects virtually to zero.

The reason the company took so long to diagnose the problem was twofold. First, they didn't understand the concept of robustness. And second, they didn't understand that problems might be caused by the interaction of two or more factors. They had been trained traditionally to vary only one at a time. Outside help taught them to solve problems where factors interacted.

The cause of these variations will be that your product lacks robustness. But you can do something about it, and the concept optimisation phase is the best time to do it.

The product's lack of performance will be due to variations in certain parameters. As yet, you will not know which they are. These design parameters are the factors that need control. To identify and put controls on them, you work through the problem systematically:

- Systematically find which design parameters are critical to the product achieving its performance. First, consider all the factors that could produce variation, and rank their probability. Then use Taguchi-type designed experiments on trial products or components to verify the key factors, and to set their tolerances. If you don't recognise this, you will produce products that vary their performance outside the limits that are acceptable to the customer. And you won't understand why because the importance of all tolerances will have equal ranking in your mind. You won't realise that maybe only one in a thousand need critical control.
- Set the economic optimum for each key parameter by tolerance design.

At the same time, plan how the product will be manufactured:

- Set economic levels of the most important processes that are needed to achieve the required accuracy and repeatability of the key tolerances.
- And finally, set the on-line control levels needed.

Many processes are affected by more than one variable, and in many cases the variables interact. So if you just vary one at a time, you will never find the optimum combination. In practical terms Taguchi-type designed experiments are the quickest way to arrive at the best values or settings for a design or process.

It needs a detailed methodology to achieve the results; it's not just gut feel or 'common sense'. You won't discover where factors interact by doing tests that vary only one at a time. Many problems have two or three interacting variables. The most we ever found was six, in manufacturing golf balls.

Taguchi-type designed experiments require the least possible time and resource to complete, while at the same time making sure that every single factor of significance is fully explored. They produce results that are virtually impossible to achieve as quickly by any other method.

Clausing[5] and Pugh give detail on the subject.

11.17 Computer tools

Computerised development tools can save huge amounts of time at the level of detailed design tasks. This is especially so on simulating the product rather than just drawing it. It is a continually expanding, inexhaustible subject. Most engineering weeklies and monthlies give extensive coverage to the latest developments. Here is a brief outline for those who may not yet have delved too far.

Computer-aided draughting software has evolved through a number of developments from 2D-drafting tools, to wire-frame and surface modelling, to solid modelling, to parametric solid modelling – what next.

[5] *Total Quality Development*, Clausing; *Total Design*, Pugh.

Then there are computer simulation tools you can use to prove out various aspects of fitness for function.

These tools have had as much impact on the process of designing as on the product's design itself. They are a major contributor to reducing the time to develop products by accurate simulation, optimising designs in a way that was not possible previously, and reducing the need for testing.

11.17.1 Computer-aided drafting

This is the first and most obvious means of saving time and effort. The CAD software tools available today permit companies to save time by automating elements of the design and engineering process. Embedding engineering rules and making the CAD database re-usable results in better-engineered products. It also shrinks time to market.

To be successful, CAD software needs the existing design engineering and manufacturing processes to be documented. They then need to be simplified in a computer model that will produce repeatable results. As the product development process continually evolves, the computer model is updated.

The more advanced CAD software is capable of reaching beyond the scope of these initial requirements. It is possible not only to document the processes involved and to embed engineering information, but also to iterate with marketing and the design studio. Alternative design studies can be generated before they ask 'what if'. This reduces time and cost compared to doing it after developing the first variant.

This method of creating a product layout at the earliest phase of development requires a sophisticated CAD model. The model establishes the parameters of every aspect of the product, from overall concept down to innermost detail.

The interface between components is developed next. Within this kind of hierarchical model, each level that is defined is more accurate and precise. Dimensional relationships between systems are defined and carried down through the model.

A single CAD model can be used to develop multiple variants of the product. The model includes parameters that will trigger specific configurations of the product being developed. This eliminates errors; for example, it will allow an engineer designing a component for variant A to refer to other elements specific to the design of the same component for variant B.

A snapshot of any of the alternative designs, even those in progress, can be available at any time. You can study the ramifications of building product variants that use common components by mixing the sub-assemblies of alternative designs.

Packages are available for PCs that are affordable by the smallest companies. Their usefulness is not just in reducing drafting time but in enabling an order of magnitude better presentation of documents for other purposes and by other departments such as quotations and photo-realistic representation to help achieve sales. Once in the machine, drawings can subsequently be modified very quickly compared to pencil and film; they can also be far more easily used as the starting point to generate another component.

Different levels of assistance can be sought from this technology. At the lowest level of technology, the product can be designed be generating only 2D drawings. These are so basic that some are given away free. But you get what you pay for. Or the product can be designed entirely in 3D, with dimensioned 2D drawings generated if required for suppliers and internal departments. Wire-frame, surface (e.g. CadKey) or solid modeller (e.g. AutoCad) programs are commonly used, ultimately to give photo-realistic renderings. At the other extreme, the design can be entered only as mathematical parameters.

11.17.2 Parametric CAD in the design process

Parametric design uses computer assistance to design the product. You define parameters and dimensions as a series of algebraic relationships that are then manipulated by the computer into the product geometry. The program uses numerous dimensional rules and logic that have been previously entered. But if a single assumption or rule is incorrect you can get nonsensical output.

11.17.3 'Knowledge-based' systems ('intelligent' CAD)

Intelligent CAD (ICAD) is a further development of parametric CAD to propose alternative design solutions from previously entered rules. It can generate design geometry that accounts for non-geometric 'rules' for example on cost, or safety requirements, that are entered into the system. It improves reliability of the finished item, improves product quality to the user and reduces investment to make it. For example Ford's C_3P system can integrate product information management data, CAD, CAE and CAM into a global system of common data. The information is also accessible to their suppliers to enable them to improve their lead-time cost and quality. It feeds previous experience from previous product development teams into the database.

> From the initiation of a project, products can change continually. This creates an interesting challenge, especially in the earliest phases.

Concentra in the USA originated ICAD in the mid-1980s; among others, Jaguar Cars, Lotus and Boeing use it. KTI Ltd's *ICAD* is one such system, Catia's *Knowledgeware* and Unigraphics' *Knowledge Fusion* are others. Computer Aided Design Inc.'s *DADS* is used by BMW and Daimler-Chrysler. Dassault Systèm's Knowledge Technologies International has a generative model system.

It saves time – in some cases a complete detail design can be produced in only 2% of your normal design time. Empirical results can be entered into the database, as well as complex dimensional constraints, such as for mating parts, and best practice design rules. The principle is to link detail and its variants to top-level functional inputs. You alter the top-level requirements and the system computes the result, so you can make fast iterations. It has reduced the time to compute a car's wheel clearance envelope from three days to an hour; suggested eight possible configurations for a bonnet inner panel in 2 hours rather than a month; designed a complete six cylinder crankshaft in 2 hours rather than a week, including the torsional mode requirements, fatigue strength and bearing loads.

But as in parametric design, you have to be sure that all the assumptions are correct; it's not a tool for those who don't understand the fundamentals of what the program is aiming at doing.

Updates on this subject are published regularly in journals such as the SAE Journal, Automotive Engineer, Engineering and so on.

11.17.4 Computer-aided engineering

Many CAD packages link into computer-aided engineering packages (e.g. ProEngineer, AutoDesk). The more facilities they give you, the more they cost. But fast payback can justify their cost. This comes because they cut time and cost in product development.

They permit the performance of the product to be modelled before testing any hardware. Tests are used to validate a model that can then be developed further to optimise the design. It is inexpensive to do this optimisation in the computer compared to repeatedly re-designing the hardware, making and testing it. The product will be more optimum, with fewer errors that would otherwise cost you a lot of money to rectify later. Using such techniques over two decades, the motor industry has halved the weight of a typical bodyshell and at the same time quadrupled its stiffness.

You can simulate a model, a process or components to compute stress levels, fluid flow, temperature profiles and so on. But beware: none substitutes for a proper understanding of the individual technique. In an absence of such understanding, companies can sometimes make serious errors from incorporating unwitting erroneous minor assumptions that only come to light when the components are tested or, worse still, used in service.

Such systems, by their accuracy of simulation, identify problems at the concept stage without the need to make and test physical prototypes. Martin-Baker, the aircraft ejection seat manufacturer, simulates crew ejection with such systems. They can optimise the design and assure the successful outcome of their ultimate test. This saves hundreds of thousands of dollars and speeds development.

Carmakers use such systems to simulate crash tests. They can simulate substituting components, or modifying them, to arrive at the best outcome before performing the actual test.

11.17.5 Further advice

An advisory body, funded by software manufacturers, that can give advice on CAD-CAM software in the United Kingdom is

The IT Centre, Computing Suppliers Federation,
8 Canalside, Lowesmoore Wharf,
Worcester WR1 2RR.
Phone – (UK) 01905 613 236

11.17.6 Manufacturing process simulation

This can be of help during the design stage to ensure that investment decisions are soundly based and that products can actually be produced as planned. There are numerous softwares for areas such as process and methods planning to speed conceptual and later detail work.

Large companies (e.g. Ford Motor Company) use advanced software to check manufacturability of new designs on their plant. For example, they discovered that a proposed concept of front rail design (B-segment car) was incompatible with the plant intended to make it. This early discovery saved them an estimated £20–40 million in costs they would otherwise have incurred to rectify the problem.

11.17.7 Project control

Programs can be obtained to 'assist' project control. But beware: in simple projects they can sometimes consume more time than they save. They can also make the process invisible within the PC and hide information from those who should be using it. Print out the resulting GANNT charts and put them on the wall in the product development team area where everyone can see them!

It is difficult to manage and monitor a process that is invisible; the purpose of technology is to help not to conceal. To control any process effectively requires that it be simplified and made visible to all those working on it. So you make sure that the GANNT charts are where everyone can see and use them; teams must be able to recognise and continuously review when targets and deliverables are required. Only if everyone has an understanding of time-scales, deadlines and tasks, will they all work to the same aims and objectives. It is usually worthwhile to use a computer to set out critical time paths and activity networks only if the project is more complex than a simple manual computation can produce.

Examples of project management software:

PrimaVera suretrack

OpenPlan

MicroPlanner X-Pert – Microplanning International (near Bristol, UK). Used by Alstom projects division to control complex engineering projects. You can plan direct onto a network diagram.

Artemis – used on North Sea platform projects.

Cascade – Mantix Ltd (Ipswich). Very sophisticated; used by large organisations with complex products, such as Rolls Royce aero-engines.

There are many more.

The base program that many use is Microsoft *Project*. But for some applications this may not have enough features. Using *InTime* with *Project* allows faster time-sheet data input. Examples of programs with different features are shown on the left.

11.18 'Design for X'

This is a generalised category of tools and techniques where 'X' stands for manufacture, assembly, reliability – whatever. The most often referred to are as follows.

11.18.1 Design for manufacture and assembly (DFMA)

If your company is still organised by department rather than by project, you will be generating significant additional cost and delay. The larger the company grows the more departmental loyalties will hinder information flow. Small companies tend to do it naturally because one person may design and production-engineer the whole product. In some larger companies, production engineers are not involved in the design process at all and it is not uncommon for the production-engineering department to re-design the product after the design department has signed it off. The production department may even do a further re-design beyond that, so that three separate sets of drawings exist within the company. While this may ultimately reduce the product's apparent manufacture cost, it wastes untold time and money, and ensures that the design team will never learn from experience or ever produce an optimum product design. In extreme cases, it can torpedo sales by the productionising process ruining customer appeal and reducing product performance.

Using multi-function project teams avoids this waste. See *Chapter 5* and section 11.4.

Design for manufacture and assembly is a methodology brought to prominence by Boothroyd and Dewhurst. You should already be project-based, so your production engineering should already be integrated with your design and marketing functions in each team developing a new product. Shared advice and information results in parts being designed so they are easier, faster and cheaper to make and assemble, and which at the same time better satisfy what the customer wants.

Boothroyd and Dewhurst advocate formal methodologies to cut manufacture time and cost. Their organisation can be contacted via Design IV at 01873 855 700 or e-mail info@dfma.com.

11.19 Low-cost tooling and models

You can produce look-alikes for people to examine and handle. Or you can make low-cost tools to make a small production run of the product. There are a variety of methods, to produce parts that have properties representative of your ultimate parts off production tools, all at low cost.

Each has its particular application. Journals such as *Engineering* are a continuous source of articles on the subject. The Materials Information Service at the Institute of Materials (UK, 0207 839 4071) publishes a review of information on low cost tooling.

You can make low cost tooling to manufacture a pilot run off tools. There are a variety of methods, from using reinforced resins, sprayed metal moulds or machined aluminium soft tools for injection moulding; to concrete-backed electro-deposited tools for large press-formed glass fibre reinforced mouldings.

You can cast structural look-alikes from vacuum cast thermoset resins using silicone rubber moulds. There are a number of routes to making an accurate mould. You can use wood, styrene foam or plaster to make the pattern model. Or you can make a stereo lithography laser cured resin model from your 3D-CAD files. Centres offering such services are becoming commonplace.

If you just want an appearance-only look-alike, there are a number of ways you can make product models quickly using tried and tested traditional mock-ups such as from structural plastics foam, body filler and sprayed paint.

11.20 Rapid prototyping

11.20.1 Understand the processes involved

Not everyone understands all the processes involved in rapid prototyping. But sufficient staff have to understand the process and why it must be done as planned. You produce a stereo lithographic (SLA) model by a dedicated machine using your 3-D CAD model to produce a solid artefact. There are a number of SLA-related processes, producing artefacts in sintered plastics, sintered metal, solid resins, paper laminates and so on.

Companies often pay to have SLA models made without understanding what they are buying or why. Some have been so enthusiastic about the sexiness of SLA, and determined to use it, that they have used the wrong processes for their project. For example some have used it to make rectangular boxes; traditional model-making techniques are both faster and cheaper, and probably more accurate. And it may not be suitable for small and delicate SLA parts because they can suffer post-production heat distortion and need to be handled very carefully. In other words, it is not a universal panacea; you need to understand what you are doing.

11.20.2 Troubles with rapid prototypes

Rapid prototyping is a key process in reducing time to market. But lack of experience or knowledge can cause problems.

Contrary to some claims, SLA models are not 100% accurate. They may look visually complete, but they generally require hand finishing, which can be key to the ultimate quality of the delivered part. So you may need to hand-finish the SLA product before you can use it to make your silicone rubber moulds for example. You may well need prototypes for market testing; but they will need considerable work before they can be used. Marketing staff are often surprised that the SLA 'prototypes' they receive are not instantly ready for their purposes.

Sometimes you need iterations; one may not suffice. And make sure that it's used by the staff who need it. It may be an interesting toy to marketing staff whereas it can be an essential tool for design staff to complete their work. Not having access to it immediately can impede the programme.

How not to, as reported at a conference: An American company had completed 2D-CAD-based production data: but their Hong Kong tool maker couldn't use 3D data, and needed 2D-CAD information and paper drawings. The American company needed a fabricated prototype to prove the production component design. The tooling needed to proceed quickly to avoid any delays from the Chinese New Year.

Fabricated prototypes would take too long, so they asked the company's model maker to turn the 2D information into 3D model files and create stereo lithography models to prove their fit and function. The tool maker was told to go-ahead with the tooling because the shorter lead-time from this route would reveal problems early enough for errors to be corrected.

But the company discovered a major problem, panicked and stopped the tooling. Because this happened over the Christmas break they couldn't sort it out until January. The models of the components wouldn't mate. Looking at the drawings it was difficult to understand why. But the model maker, in turning the 2D data into 3D model files, inadvertently created two left-hand components instead of a left and a right.

By that time a conventional route could have fabricated a model. The tooling was delayed three weeks, and a cost saving turned into an increase in cost and a delay. Always understand what you're doing.

11.21 Engineering data sources

There are many catalogues available from manufacturers on CD-ROM that are more conveniently stored than paper volumes. Some companies have them on-line in their company Intranet so everyone can access them. Companies such as Findlays Publications publish comprehensive collections of such data, also available through the Internet (e.g. www.designselectorglobal.co.uk). These allow searches for suppliers of engineering components, materials and services.

Some organisations, such as the UK Spring Research and Manufacturers Association (SRAMA), issue free programs on disc to perform engineering calculations on, for example, spring design.

11.22 Identifying process waste

This is all about discovering where you spend time on activities that contribute nothing to what the customer vales, but nonetheless pays for. In other words, how might you cut cost out of the process? There are two tools you could consider, value stream mapping and design structure matrix analysis

First of all, to eliminate misunderstanding that is creeping into our buzzword vocabulary, because many companies have been learning to apply 'lean' processes to their production area, some clarification. 'Lean' means using Toyota production system (TPS) methodology. Metrics are annual stock turn and order lead-time. You are not lean is you have only applied 6-σ or 5-s; those are helpful discipline techniques, not the fundamentals of lean. Lean means single piece flow, single minute set-ups, cells not lines, making to order not to forecast and other TPS attributes. (See Womack and Jones' *Lean Thinking*.) Well-implemented TPS can make staggering reductions in manufacture cost, floor space and inventory with big quality improvements. For example, for a £15,000 outlay, one Midlands company was (2004) saving £100,000 a year from one line alone, and, having learned how to do it, were spreading it across their whole production operation. Their target was to cut £500,000 direct cost plus £1 million bank borrowing out of £20 million turnover. If you haven't done it, your manufacturing cost base could be 30% greater than it could be. It's essential for long-term survival in UK manufacturing.

If you're applying TPS methodology in your production area, you will almost certainly be value stream mapping (VSM) your production processes. Because the benefits are so impressive, it's useful to understand how you might apply VSM your product development process.

There are significant differences between a production process, or a routine administration task such as order processing, and developing a product. In the two former processes, you can see the product, and observe and map how it is moved, stored, worked on – and so on. Whereas in product development you can see only what is happening at this instant in time. You can recall what you did before, and what you did after what you're doing now, by remembering how you developed another product. So you can map your process to identify redundant tasks, unnecessary delays and so on. That is useful.

11.22.1 What is the PD equivalent to the Toyota production process?

There is less information on how you value stream map (VSM) the product development process than on manufacturing. The original work by Shingo and Ohno related to production systems, as does Womack and Jones' *Lean Thinking*. Toyota does use a smart product development process: they have a product development culture and attributes that few Western companies can emulate. It's more than just a good process. As well as their culture, and an organisation larger than yours, their chief engineers are key: they apply unusually wide detail knowledge from 20 to 30 years of working on the detail of the product building blocks and systems prior to promotion.

It is certainly possible to VSM your product development process. The seven classic wastes identified by Shingo and Ohno have product development equivalents. An ex-Toyota member recently explained how a Ford US tool and die supplier has run their version for just the design process (courtesy of Dan Jones' UK Lean Enterprise Academy, see www.leanuk.org). But it used a different set of VSM symbols and protocols from those for production operation. It identified waste effectively; but you won't replicate how Toyota does it based only on VSM analysis.

11.22.2 Design structure matrix analysis

Another tool is design structure matrix (DSM) analysis devised by professor Eppinger of MIT. This analyses how to re-order the tasks you perform in the sequence that wastes least effort. You discover all the interactions between your product development tasks such as which tasks need information from which others. This demonstrates the best order to do them in. Because communication interdependencies are identified, you can also discover which people should be in which groups to run a project. Large complex organisations have gained benefit from applying the methodology and it's certainly one you should know about. It's useful in a different kind of way from value stream mapping.

See www.dsmweb.org/index.html for more information.

11.22.3 Caveats

Like all the tools and techniques mentioned above, there are caveats to both VSM and DSM. They won't necessarily reveal the vital something that is missing from how you run your product development because the exercise is limited by your own experience. Indeed, the same applies to mapping your production operation. You could redesign your processes to make them smarter, which would alter your process map. With production operations and processes however, your plant and your planned process route govern it far more, and you are more likely to know how you should be operating, partly because of the vast sea of continually updated published information. There is not the same situation in product development; there isn't much 'fixed plant' and the process is limited more by your intellectual invention. It has been said that great process kaizen can't make up for lousy product design.

> Unfortunately, no combination of techniques will solve all your problems. It relies on culture as well as process.

So if you don't know that some tasks should be done, or how to do them, or even that they exist, analysing what you've done in past projects may not reveal the best way to do it. You may never have done certain tasks in your product development process at all; or diagnosed that you should. I've been fortunate in being able to analyse in depth how more than 80 really profitable product developers do it, and no two are alike. Nor, within each, are two projects alike. But by detailed analysis and, perhaps more important, having been there and done it, it was possible to separate the useful from the less so. This led to the discovery that there is a common underlying baseline blueprint that fit all of them. This is what *Chapter 9* sets out. But, interestingly, even most of the best don't do all of it – probably doing 80% of it is enough to make them really successful.

How Toyota do it is summarised in 9.4.6; it's about culture and method, not VSM or DSM. But, if just using the right tools were all that is needed, many companies would be as competent at it as Toyota. But they're not.

So it is certainly useful to know how to VSM and DSM your current process. But before you embark on either, put the methodology into action that is detailed in *Chapter 9*. It gives you at least as much if not more benefit before you apply the analysis. You can learn and apply the analysis to identify further waste but only after you operate your system well.

Chapter 12
Bibliography

This book intentionally omits source detail that is well described in other works. But which works are useful? Which are the best to use when you want your staff to learn the detail to apply to your programmes?

The following, referred to in the text, form a concise library that give guidance on detail operating aspects of the process. The international standard book number (ISBN) enables your bookseller to identify it easily when ordering. Where shown, order it direct by phone from the publisher or agent, because it may not arrive through your local bookseller at all!

D. Allen, *Developing Successful New Products*,
FT/Pitman UK, ISBN 0-273-60150-4
(out of print but available from wp@mynott.com, £80, $120)
The definitive handbook on product planning, by a senior manager who spent his life doing it – excellent account of experience and lessons learned.

M.F. Ashby, *Materials Selection in Mechanical Design*,
Pergamon; ISBN 0-08-041907-0
A fascinating handbook that will change you perception of how you should go about selecting the best material for any application. Goes right back to basics. Every company needs this.

R.M. Belbin, *Management teams – why they succeed or fail*,
Heinemann, ISBN 434-90126-1
A descriptions of why and how teams work and the mix of personalities that you need in the team to achieve the best performance.

J. Bicheno, *The lean toolbox*,
Picsie Books, ISBN 0-9513-829-9-3
The reference guide to the tools and techniques of lean operation and lean enterprise. Excellent read-cross to how you organise product development – an essential reference.

D. Clausing, *Total quality development*,
ASME Press, ISBN 0-7918-0035-0 (UK 01462 437 933)
A manual on many of the detail aspects of the process of developing products. (Don Clausing is Professor at MIT.) Especially useful on operating QFD, investigating customers and optimising costs and concepts.

Cranfield University UK, *Using concurrent engineering for better product development*, 1999, ISBN 1-87131-575-1
(out of print but available from wp@mynott.com, £95, $140)
A workbook developed by installing the techniques in several companies. You will resolve all your turf wars with full team working across all functions. If you have problems installing team-working, this will fix them using the lessons learned by others who faced and overcame theirs.

M.A. Cusumano and K. Nobeoka, *Thinking beyond lean*,
The Free Press, ISBN 0-684-84918-6
How the largest automobile manufacturers run their product development process with multiple projects; compares a number of systems with how Toyota does it. Interesting lessons that non-auto companies could adopt.

P. Lencioni, *The five disfunctions of a team*,
Wiley, ISBN 0-7879-6075-6
A readable guide to how teams behave in practice and how you might reshape their members' attitude to achieve star performance.

C. Mynott, *The product development process – an outline blueprint*,
ISBN 0-9538779-0-7
(available only from wp@mynott.com, £150, $300)
A summary of the process with further background, and templates and checklists to aid running and control of the process – complementary to this book.

C. Mynott, *Product development case studies*, ISBN 0-9538779-0-8
(available only from wp@mynott.com, £150, $300)
More than fifty case studies illustrate how companies went about developing new and improved products and product ranges and the lessons they learned. In various sectors, from small to major internationals. Continually updated.

P. O'Connor, *Practical Reliability Analysis*,
Wiley, ISBN 0-471-90551-8
Handbook on a number of useful techniques on analysing product reliability.

G. Pahl and W. Beitz, *Engineering Design*, Springer-Verlag, ISBN 3-540-50442-7
Useful for developing the engineering detail of products.

S. Pugh, *Total Design*, Addison Wesley; ISBN 0-201-41639-5
Useful for design principles of developing products.

D.G. Reinertsen, *Managing the design factory*, 1997,
The Free Press, ISBN 0-684-83991-1
How you apply vital tools to the development process, including creating a economic models, why facility queues occur and how you deal with them, product architecture, information from failures, risk, metrics and so on.

M. Rother and J. Shook, *Learning to see*, 1998, the Lean Enterprise Institute, ISBN 0-9667843-0-8 (USA 001 617 713 2900)

A workbook on how you analyse value streams to help identify wasted effort, to enable you to target cost reduction. Can be applied to any process, whether production or administrative. A unique work.

S. Shingo, *Non Stock Production*, 1988, Productivity Press, ISBN 0-915299-30-5 (UK 01462 437 933)

A treatise on how you put in place what has come to be known as the Toyota Production System; the general bent to logical thinking and problem solving is of particular interest to product developers.

S. Shingo, *A study of the Toyota production system*, 1981, Productivity Press, ISBN 0-915299-17-8 (UK 01462 437 933)

A description of the Toyota Production System from a production engineering point of view; an excellent source work on exactly what it is and how it operates. Dispels misleading impressions that have emerged over the past decades.

P. Smith and D. Reinertsen, *Developing products in half the time*, 1991 Van Nostrand, ISBN 0-442-00243-2

Some key management techniques that improve project profit, especially how you model projects to understand what most influences their success.

J.P. Womack and D.T. Jones, *Lean Thinking*, 1996, Simon and Shuster, ISBN 0-684-81035-2

A description of how companies that have applied the Toyota production system and related good practice have radically cut their costs and gained market share.

J.P. Womack, D.T. Jones and D. Roos, *The Machine that Changed the World*, 1990, Simon and Shuster, ISBN 0-89256-350-8

The results of a five-year world-wide programme based at MIT comparing the operating practices of carmakers, showing how the Toyota production system leads the way in commercial success.

Index